Science

For further volumes:
http://www.springer.com/series/11657

Science and Fiction – A Springer Series

This collection of entertaining and thought-provoking books will appeal equally to science buffs, scientists and science-fiction fans. It was born out of the recognition that scientific discovery and the creation of plausible fictional scenarios are often two sides of the same coin. Each relies on an understanding of the way the world works, coupled with the imaginative ability to invent new or alternative explanations—and even other worlds. Authored by practicing scientists as well as writers of hard science fiction, these books explore and exploit the borderlands between accepted science and its fictional counterpart. Uncovering mutual influences, promoting fruitful interaction, narrating and analyzing fictional scenarios, together they serve as a reaction vessel for inspired new ideas in science, technology, and beyond.

Whether fiction, fact, or forever undecidable: the Springer Series "Science and Fiction" intends to go where no one has gone before!

Its largely non-technical books take several different approaches. Journey with their authors as they

- Indulge in science speculation—describing intriguing, plausible yet unproven ideas;
- Exploit science fiction for educational purposes and as a means of promoting critical thinking;
- Explore the interplay of science and science fiction – throughout the history of the genre and looking ahead;
- Delve into related topics including, but not limited to: science as a creative process, the limits of science, interplay of literature and knowledge;
- Tell fictional short stories built around well-defined scientific ideas, with a supplement summarizing the science underlying the plot.

Readers can look forward to a broad range of topics, as intriguing as they are important. Here just a few by way of illustration:

- Time travel, superluminal travel, wormholes, teleportation
- Extraterrestrial intelligence and alien civilizations
- Artificial intelligence, planetary brains, the universe as a computer, simulated worlds
- Non-anthropocentric viewpoints
- Synthetic biology, genetic engineering, developing nanotechnologies
- Eco/infrastructure/meteorite-impact disaster scenarios
- Future scenarios, transhumanism, posthumanism, intelligence explosion
- Virtual worlds, cyberspace dramas
- Consciousness and mind manipulation

Paul J. Nahin

Holy Sci-Fi!

Where Science Fiction and Religion Intersect

Springer

Paul J. Nahin
Department of Electrical & Computer Engineering
University of New Hampshire
Durham, New Hampshire
USA

ISSN 2197-1188   ISSN 2197-1196 (electronic)
ISBN 978-1-4939-0617-8   ISBN 978-1-4939-0618-5 (eBook)
DOI 10.1007/978-1-4939-0618-5
Springer New York Heidelberg Dordrecht London

Library of Congress Control Number: 2014932675

Printed on acid-free paper

Springer is part of Springer Science+Business Media (www.springer.com)

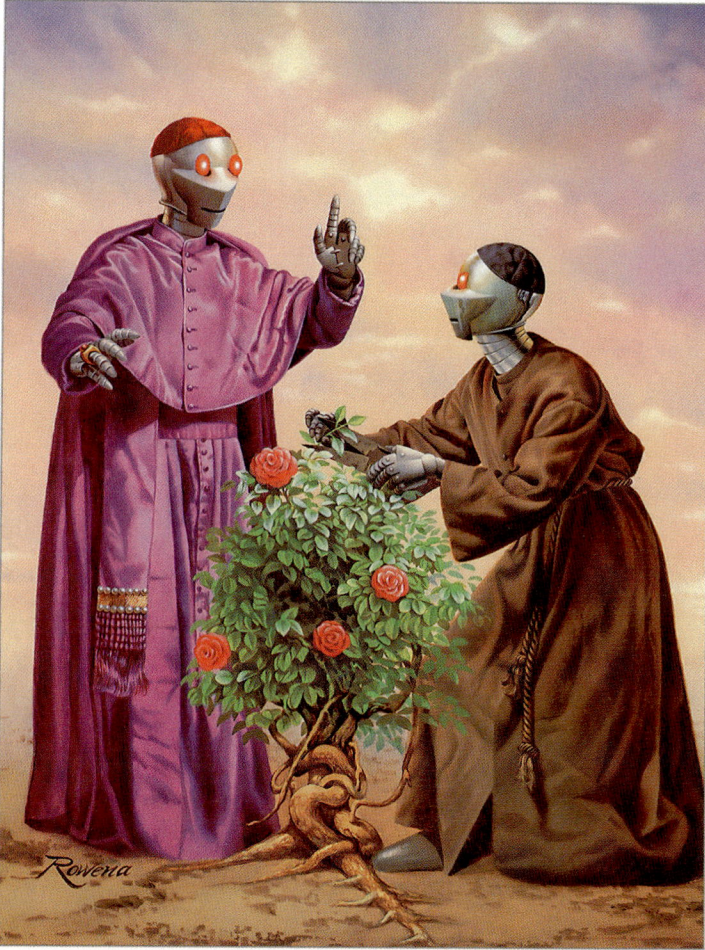

Frontispiece illustration reproduced by permission of the artist, Rowena Morrill. Two robotic priests tend a rose bush in Clifford Simak's 1981 novel *Project Pope*. Located on the remote planet End of Nothing at the far edge of the galaxy, the colony outpost called Vatican-17 has the goal of creating a universal religion led by an immortal computer Pope.

Also By Paul J. Nahin

Oliver Heaviside (1988, 2002)
Time Machines (1993, 1999)
The Science of Radio (1996, 2001)
An Imaginary Tale (1998, 2007, 2010)
Duelling Idiots (2000, 2002)
When Least Is Best (2004, 2007)
Dr. Euler's Fabulous Formula (2006, 2011)
Chases and Escapes (2007, 2012)
Digital Dice (2008, 2013)
Mrs. Perkins's Electric Quilt (2009)
Time Travel (1997, 2011)
Number-Crunching (2011)
The Logician and the Engineer (2013)
Will You Be Alive Ten Years From Now? (2014)

*For Patricia Ann who, for reasons known only to God, Himself,
always had faith in me.*

Epigram

"Science fiction bears the same relation to the world of science and technology that legends of the saints do to the Christian religion."
    –John Robinson Pierce, in *Engineering & Science* (November 1981)[1]

---

[1] J. R. Pierce (1910–2002) was a 1936 Caltech PhD in electrical engineering, was executive director of research of the Communications Sciences Division at Bell Labs (where he gave the transistor its name and proposed the Echo 1, Telstar, and Relay communication satellites), and was Chief Technologist at Caltech's world-famous Jet Propulsion Laboratory. Not so well known is that in March 1930 he published his first science fiction story—"Relics from the Earth"—in the Hugo Gernsback pulp magazine *Science Wonder Stories*. Two dozen more tales and essays about the future followed over the years, in magazines ranging from the elite *Magazine of Fantasy and Science Fiction* to the more worldly—and far better paying—*Playboy* and *Penthouse*. The above quotation is from an interview I did with him for the Caltech alumni magazine (http://calteches.library.caltech.edu/527/2/Nahin.pdf).

Note to the Reader

There are many religions in the world, but the stories discussed in this book mostly assume either Christianity or something "vaguely Christian." If a priest is a character he is often a Catholic, and then usually he is a scholarly Jesuit, almost always appearing as a mathematician, a biologist, or a physicist. Typical is the Jesuit scientist-hero in the 1958 novel *A Case of Conscience* who, when faced with an enormous moral dilemma (on a planet 50 light-years from Rome) that intersects science and religion, is described as follows: "a lifetime of meditation . . . had made [him], like most other gifted members of his order, quick to find his way to a decision through all but the most complicated of ethical labyrinths. All Catholics must be devout; but a Jesuit must be, in addition, agile."

A good, that is, *interesting*, science fiction story requires emotional tension, and that is the natural result in having a member of the Society of Jesus caught in conflict between his spiritual faith and his scientific intellect. This isn't *always* the case in science fiction, however—in one story Tibetan monks appear, in another we read of a rabbi, and in yet another we encounter the spirit world of American Indians—but it is pretty nearly the case. This isn't an intentional snub of other religions, but simply recognition of the fact that most science fiction writers in the English language were raised in a Christian-based culture, even if they themselves are/were not Christians. After all, you write best, the old saying goes, about what you know. When it comes to the religions of alien beings from outer space, however, well then, of course all bets are off!

I have also limited my presentation to stories that have appeared in English, either directly or in translation (for example, the novels of Stanislaw Lem, originally published in Polish). This isn't to deny the fact that non-English science fiction writers have had a lot to say in their tales about religion, but rather it is simply that my linguistic skills are limited! Spanish, in particular, has been the language of numerous stories dealing with what might happen if the Catholic Church ever encounters alien theology.[2]

I have limited myself to discussing science fiction in the written word and mention movies only in passing (and TV stuff not at all). I made that decision because movies are often based on previously published *written* works, and the transition process from one medium to the other has usually resulted in a decline of merit. The "Hollywood effect" is, more often than not, not a particularly good one for science fiction.[3] And the less said about TV science fiction the better.

That last sentence is a pretty damning claim, and, like most blanket statements, there *are* exceptions. It is generally agreed in the science fiction community that two episodes of the 1964 season of TV's *The Outer Limits* ("Soldier" and "Demon with a Glass Hand") and an episode in 1967 on *Star Trek* ("The City at the Edge of Forever") were pretty darn good science fiction. On the other hand, all three works were the work of a *single* writer—Harlan Ellison (born 1934)—and that sad fact (along with all three works dating from a half-century ago!) lends support to my harsh assessment.

---

[2] See Elizabeth Small, "Religious Institutions in Spanish Science Fiction," *Science-Fiction Studies*, March 2001, pp. 33–48. In one of the tales mentioned in this essay, "El orgullo de Dios" ("The Pride of God") by Pedro Jorge Romero (published in 2000), the Church engages in direct physical combat with Satan and "the Earth is vaporized, and a militarized Catholicism spreads through the galaxy, with monasteries and convents as the front lines of defense." If you think that this is just "science fiction talk," consider these words by General of the Army Douglas MacArthur, in a speech to the cadets of the US Military Academy in May 1962: "We deal now, not with things of this world alone, but with the illimitable distances and as yet unfathomed mysteries of the universe. We are reaching out for a new and boundless frontier. We speak . . . of spaceships . . . of ultimate conflict between a united human race and the sinister forces of some other planetary galaxy . . . ."

[3] A funny, insightful essay on the mostly dismal record of science fiction in the movies is "The Imagination of Disaster" by Susan Sontag (1933–2004) in her collection *Against Interpretation*, Farrar, Straus & Giroux 1966, pp. 209–225. This essay appeared before the 1968 release of *2001: A Space Odyssey*, however, and I think Sontag would have had better things to say about that film as well as such romantic or funny SF films as *Somewhere in Time* (1980), *Back to the Future* (1985), and *Bill & Ted's Excellent Adventure* (1989).

Acknowledgements

Holy Sci-Fi!, with two exceptions, is very different from all the other books I have written. But even those two exceptions, treating the physics of time machines and the paradoxes of time travel as presented in science fiction, avoided considerations of the human soul, a central issue in this book. So, even among those who were supportive of my books on mathematics, physics, and electronics, there was surprise and more than a little caution when I revealed my new project. Still, there were three people who *did* greet the writing of *Holy Sci-Fi!* with immediate enthusiasm.

The well-known SF writer (and professor emeritus of physics at UC/Irvine), Gregory Benford, liked the idea a lot and generously allowed me to reprint two of his short stories that originally appeared in *Nature*.

Trevor Lipscombe, my former math editor at two university presses (and now Director of a third, well-known university press dealing with both medieval and modern theology), also liked the book—but felt it perhaps not quite right for the "serious theologians" that make up most (if not all) of his present audience. Trevor, almost certainly I have to admit, will not agree with everything I say in *Holy Sci-Fi!*, but I know that rather than being upset with me he will instead sincerely pray for the salvation of my soul.

My wife, Patricia Ann, to whom I've dedicated this book, is the only reason I am still alive to write it. Without her I would long ago have eaten far too many pepperoni pizzas, french fries, and apple pies and would by now have personal knowledge of the truth about Heaven (or, at least as likely perhaps, of Hell).

Finally, I must thank the great people at Springer whose support made all the difference: Jace Harker (past physics editor), Amita Raval (present physics editor), and Ho Ying Fan (assistant physics editor). In addition, two anonymous reviewers gave me the benefit of a number of helpful comments after reading the original typescript, for which I am most grateful. During the final production phase of the book, Brian Halm in Springer's New York Book Production Department, and Project Manager Rekha Udaiyar of SPi Technologies in Chennai, India were a pleasure with whom to work.

Lee, NH   Paul J. Nahin
January 2014

Contents

Chapter 1
Introduction

1.1 Author's Note One

I am not a religious person, in the sense of believing in a supreme being who is the ultimate cause of the world we immediately live in, or of the universe at large in which our world is but an extremely tiny part. I am not even a deist. In other words, I am not someone who at least believes in a Creator, while not going so far as to further believe that He/She/It cares about human affairs. In fact, to be up-front about it, I confess to being an agnostic (a polite atheist). For all my readers who *are* true believers, however, please understand that I am not aggressively hostile about this issue. I don't think it silly to believe, and I am even willing to admit I could be wrong. I simply haven't been convinced that I am in error. I almost certainly don't have to discuss here the difference between being an agnostic and an atheist, but I do like the following illustration of an agnostic, an atheist, and a true believer:

> *True Believer: God made the heavens and the Earth.*
>
> *Agnostic: Prove it.*
>
> *Atheist: There is no way that God exists.*
>
> *Agnostic: Prove it.*

None of the above means that I don't find it a glorious event when I see a rainbow in the sky. Instead of creating a 'toasting marshmallows over a campfire' tale about dancing elves in green pants and pots of gold being the reason for that wondrous vision, however, or some other equally fanciful 'explanation,' I look for a rational underpinning to the colorful arc in the laws of physics and the rules of mathematics.[1]

---

[1] For more on the mathematical physics of the rainbow, see my book *When Least Is Best*, Princeton 2004 (corrected paperback 2007), pp. 179–198.

P.J. Nahin, *Holy Sci-Fi!*, Science and Fiction,
DOI 10.1007/978-1-4939-0618-5_1, © Springer Science+Business Media New York 2014

For some, any mention of physics and math brings back unpleasant memories of Mr. Scienceguy's boring high school class (I know *you* aren't in this category!) , along with the feeling that technical subjects somehow lack the compassion, the understanding and forgiveness, the *loving comfort* of an all-forgiving God. The world is undeniably a harsh place, and the concept of God offers an emotional refuge from what would otherwise simply be a mean and brutal existence from birth to death. To the lower animals the universe may well be, as Tennyson wrote, "red in tooth and claw," but for creatures with souls (as so many believe are the unique possession of humans) there just *must* be something beyond the dry, pitiless, morality-neutral laws of math and physics. Or so do many believe.

One person who would surely have felt that way was the famed essayist Charles Lamb, at the so-called "Immortal Dinner," a party given on December 28, 1817 at the home of the English painter Benjamin Haydon. In attendance at what Haydon modestly described as "a night worthy of the Elizabethan age . . . with Christ hanging over us like a vision" were such luminaries as the poets Wordsworth and Keats. That evening Lamb toasted a portrait of Isaac Newton with words describing Newton as "a fellow who believed nothing unless it was as clear as the three sides of a triangle, and who had destroyed all the poetry of the rainbow by reducing it to the prismatic colors."

Lamb was described by Haydon a having been "delightfully merry" just before he made his toast, which I suspect meant he was thoroughly drunk. Still, one of Lamb's younger dinner companions was greatly influenced by that toast, as 3 years later John Keats repeated the sentiment in his poem *Lamia*, where we find the words

" . . . Do not all charms fly
At the mere touch of cold philosophy?
There was an awful rainbow once in heaven:
We know her woof, her texture; she is given
In the dull catalogue of common things.
Philosophy will clip an Angel's wings,
Conquer all mysteries by rule and line,
Empty the haunted air . . .
Unweave a rainbow . . ."

Much better, I think, and in the spirit with which I've written this book, are the following words by the English poet William Wordsworth (written in 1802, years before he attended the Immortal Dinner):

"My heart leaps up when I behold
    A rainbow in the sky;
So was it when my life began;
So it is now that I am a man;

So be it when I shall grow old;
    Or let me die!
The Child is father of the Man;
And I could wish my days to be
Bound each to each by natural piety."

In a famous 1954 science fiction story, "The Cold Equations" by Tom Godwin (1915–1980), the conflict of physics versus poetry was powerfully illustrated in a way many found to be shocking. The entire story is set in the cabin of an Emergency Dispatch Ship (EDS) ferrying a load of urgently required medical supplies to a colony on a remote planet at the frontier of the galaxy. The ship has just enough fuel to make the trip with the expected payload—if there is either just a bit less fuel *or* just a bit more payload, the EDS will fall short. Partway into the trip, the pilot discovers there is a stowaway on-board, a young girl who snuck aboard to hitch a ride to see her brother who is one of the colonists.

She knew what she had done was wrong, but had thought she'd merely be lectured, or perhaps fined. Instead, the pilot tells her she is in much deeper trouble. There is no possibility of the EDS returning to base, as it was launched into space from a hyperspace mother-ship that had briefly 'dropped into normal space' to start the EDS on its way. The mother-ship had then vanished back into hyperspace. The EDS had only one way to go, to the colony. But it couldn't make it with the stowaway on-board.

The laws of physics allowed only one solution—the payload had to be reduced. The medical supplies couldn't be touched, as without *all* of them many men would die on the colony. It was the girl that had to go. There was no other possibility, as the story tells us that

"Existence required order, and there was order; the laws of nature, irrevocable and immutable. Men could learn to use them, but men could not change them. The circumference of a circle was always pi times the diameter, and no science of man would ever make it otherwise. The combination of chemical A with chemical B under condition C invariably produced reaction D. The law of gravitation was a rigid equation, and it made no distinction between the fall of a leaf and the ponderous circling of a binary star system. . . . The laws were, and the universe moved in obedience to them. . . . The men of the frontier had long ago learned the bitter futility of cursing the forces that would destroy them, for the forces were blind and deaf; the futility of looking to the heavens for mercy, for the stars of the galaxy swung in their long, long sweep of 200 million years, as inexorably controlled as they by the laws that knew neither hatred nor compassion. The men of the frontier knew . . . h amount of fuel will not power an EDS with a mass of m plus x safely to its destination. To him [the EDS pilot] and her brother and parents she was a sweet-faced girl in her teens; to the laws of nature she was x, the unwanted factor in a cold equation."

There was no last-minute Hollywood-movie rescue to save the day; so she was ejected from the EDS. The pilot felt terrible about it, yes, but there simply was no alternative:

"A cold equation had been balanced and he was alone on the ship. Something shapeless and ugly was hurrying ahead of him ... but the empty ship still lived for a little while with the presence of the girl who had not known about the forces that killed with neither hatred nor malice."

As with so much of modern science fiction, H. G. Wells (1866–1946) anticipated Godwin on the indifference of nature to the needs of men in his 1899 short story "The Star." There we read of the approach of an enormous mass, a rogue planet from the depths of space, as it plunges into the Solar System. Colliding with Neptune, "the heat of the concussion had incontinently turned two solid globes into one vast mass of incandescence." Then, perturbed by Jupiter's gravity, this flaming new star appears to be on a collision course with Earth. A "master mathematician" who has calculated the star's new orbit declares "Man has lived in vain." But he was wrong—it's 'just' a near miss and Man survives. Indeed, to the Martian astronomers who have watched the almost-but-not-quite fatal disaster unfold from afar, little seems changed. As Wells' last sentence eerily expresses the indifference of nature, the Martians' blasé evaluation "only shows how small the vastest of human catastrophes may seem, at a distance of few million miles."

Eight decades later the science fiction author, editor, and critic Algis Budrys (1931–2008) asserted the writers that had come after Wells had learned the lesson of "The Star" well. As he wrote in one of his many erudite book review columns, "The essential thing [in modern science fiction] is the effect on human thought of the fundamental discovery that the Universe does not care; it simply works. There is no way to repeal or amend physical laws. The rich, the poor, the holy and the unholy are all subject to hunger, thirst, pain, and death. . . . And yet, how appealing it is to think that simply displaying the proper attitude might modify the Universe! It's a hope we somehow cannot bring ourselves to abandon." Budrys never mentions religion, miracles, or God in this essay, but it is difficult to believe he wasn't thinking of them when he wrote.[2]

---

[2] See Budrys' essay in the "Books" column of *The Magazine of Fantasy and Science Fiction*, May 1979, pp. 19–28. As the Polish science fiction writer, critic, and analyst Stanislaw Lem (1921–2006) wrote 2 years earlier in the same spirit as did Budrys, "it makes no sense at all to look at the universe from the

The Universe *is* a violent place. When most people think of the 'end of the world' the image of nuclear way is perhaps the first one to come to mind. Such a war would be terrible, of course, but it would be small potatoes compared to what the Universe is capable of doing to us by merely following the laws of physics. To start off 'small,' just imagine what a rock 10 miles in diameter smashing into Earth at 50,000 mph would do. Indeed, *has* done, numerous times, in the past. The last time it happened, 65 million years ago, the dinosaurs vanished forever. And on a grander scale, we of course have the scenario in Wells' "The Star." Such impacts sound pretty bad, but at least we would see them coming at us and, perhaps armed with a sufficiently advanced technology, we could even do something about it. The Universe has even worse possibilities for us, however.

Things like supernovas and gamma-ray bursters (a massive star that reaches the end of its fusion life is no longer able to support itself against gravitational contraction, and so collapses into either a neutron star or a black hole, respectively), releasing in a flash more energy than the Sun will radiate over its entire existence! If such a thing happened close to Earth (where 'close' means anything perhaps as far away as several thousand light-years) then we could literally be toast. And we'd never see it coming, as the radiation energy of such stupendous explosions travels at light speed. Just think, such a monster wave of energy could be just two light-seconds from Earth *right now* and you'll be dead before you finish this sentence. Sound like SF nonsense?

No, it isn't, and such things are happening in the Universe *right now*. It is estimated that there are a hundred billion galaxies (our own Milky Way is one) in the observable Universe, each with a hundred billion or so stars. On average, *one* of those stars in each galaxy becomes a supernova once each century. This is just an average and, in fact, the last supernova observed in the Milky Way was five centuries ago and many thousands of light-years distant. Perhaps one-a-century doesn't sound like much to you but, working with that average value, it is simple arithmetic to conclude that there are 30 supernovas *each second* somewhere in the Universe! That's a billion supernovas each year. To end life on Earth, all it would take is *one* of them to occur within, say, 2,000 light-years. We'd never see it coming—and that would perhaps be a blessing, God's final gift to us even as He (through His laws of nature) lights the match.

A more subtle treatment of Wells' idea in "The Star" that gets 'closer' to God and His laws of nature is in a 2001 tale, "Anomalies," by the physicist Gregory Benford (born 1941). Astronomers discover that the moon is suddenly out of position, too far ahead in its orbit "by several of its own

---

standpoint of ethics." See his "Cosmology and Science Fiction," *Science Fiction Studies*, July 1977, pp. 107–110.

diameters." And the tides on Earth are slightly off, too. The scientific community is perplexed until it is suspected that what has happened is a 'cosmic error' in the logical computation of the state of the universe. As one character expresses it, "God's a bloody mathematician?" Like any good computer, however, the universe has error-correction capability and the moon is soon back to where it should be. The episode does have one lasting result, though—the founding of a new scholarly field, that of *empirical theology*!

As I think Benford was hinting at, one might well argue that Godwin's 'cold' laws and rules, and Newton's 'cold' philosophy (to use Keats' word) were created by a supreme being, who thereafter remains hidden from us and simply allows everything else to 'naturally' occur in accordance with those laws and rules. And, in fact, I have no real quarrel with that viewpoint, but would simply add to it that the 'supreme being' can then only be 'known' through those laws and rules, and so it is, ultimately, only those laws and rules (what we call *Nature*) that interest me.

I agree with one of the characters in the 1981 novel *Project Pope*, by Clifford Simak (1904–1988), who becomes involved in a search for the physical location of Heaven. At one point she comes to reject the idea of "a never-never land that could exist with no need of either time or space and, presumably, *without the steadying hand of the physical laws that went with them* [my emphasis]."

One analyst who would surely have disagreed with me (and Simak) on this issue is the lay theologian (and late Professor of Medieval and Renaissance Literature at Cambridge University) C. S. Lewis (1898–1963). Lewis held that the natural laws *could* exist quite well with what he called God's occasional "interventions," if so required (for example) to bring about a miracle. For many (if not all) scientists, however, the true miracle is just the inverse of Lewis' claim: the known natural laws appear to apply *everywhere* in the observable universe, at *all* times, with *no* exceptions. The discovery of even a single violation would be a guaranteed Nobel Prize, so I don't think anyone is covering anything up! I'll leave Lewis for the moment, but we'll return to him in the next section.

I do suspect that even Lewis would have drawn the line at the thesis in the 1982 story by Hilbert Schenck (born 1926), "The Theology of Water," in which the physical properties of water differ from place-to-place in the universe, becoming whatever are required for humans to flourish in each place. On Saturn's moon, Titan, for example, water freezes not at 32 °F, but at a different temperature because that 'works out better' on Titan for humans. Such a conceit is so broad that I think even Lewis would think God had gone mad to have arranged matters for such continuous, universe-wide miracles to occur.

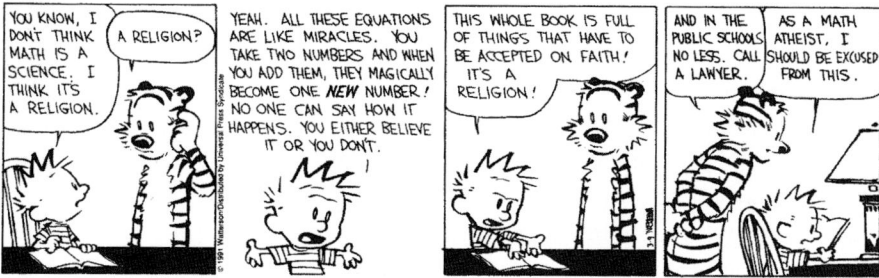

Whether or not you agree with my views is not important for deciding whether or not to read this book. I've included these comments because I know readers are generally curious about an author's personal position when his or her book is on a controversial subject—and that certainly includes religion. This is a 'what-if' book, a book in 'experimental imagination,' if you will, and not a 'you'd better believe it or you are going to go straight to Hell' book. If Rod Serling had included religious tales in his 1960s "Twilight Zone" television classic, the stories I've selected to discuss in this book could well have served as starting points for scripts. There *were* "Twilight Zone" episodes in which supernatural characters appeared; for example, Mr. Death, Mr. Fate, and yes, even the Devil, but they were the sort of tales that I think only occasionally (if at all) attempted serious theological speculation.

You don't have to believe in God to be interested in the search for God. The fact that other people *do* believe is enough to make *their* search non-trivial. I'll be even stronger on this last point: while I don't think (as do the more extreme skeptics) that it is plainly madness to believe in a supreme, personal being (God), I have to say that, *in my opinion*, it does necessitate the suspension of rational skepticism. What is ultimately required—and surely this is no surprise—is *faith*.

1.2 God and Skepticism

In a brilliantly original book,[3] the political scientist Steven Brams uses the mathematics of two-person game theory to study the outcomes of an ordinary person interacting with an 'opponent' who possesses the attributes of

---

[3] Steven J. Brams, *Superior Beings: if they exist, how would we know?*, Springer-Verlag 1983. This book is the sequel to Brams' first application of game theory to theology: *Biblical Games: a strategic analysis of stories in the Old Testament*, MIT Press 1980. Brams is professor of politics at New York University. See also Appendix 1.

omniscience, omnipotence, immortality, and incomprehensibility. That is, he studies the interaction of a human 'playing against' what he called a 'superior being' (SB)—or, in a theological setting, God. The first three attributes are clearly those that no human has ever possessed, and so can fairly be called *supernatural*. The fourth attribute of incomprehensibility, however, is *not* beyond the reach of ordinary humans.

As Brams defines the term, incomprehensibility is just the standard game theory concept of using a *mixed strategy*, which means that if a 'player' has two or more possible responses to each of his opponent's decisions, then the 'player' should choose among those various available possibilities according to some probabilistic rule. In that way the player's behavior from game-to-game will appear to others to be arbitrary. Brams makes the interesting argument that the rational use of arbitrariness may offer an explanation for what may well be the central conundrum of theology: why, even with a benevolent God, do evil things still happen? The science fiction writer Poul Anderson (1926–2001), in his 1973 story "The Problem of Pain," argued as did St. Augustin in his *Confessions*, that moral evil—defined as the willful disobedience of God—is the logical result of giving man free will, and that the *real* conundrum of theology is the question of why there is undeserved pain, such as an agonizing, prolonged death by cancer.

Another of Brams' surprising conclusions is that there exist possible inter-actions between the SB and a person in which the formidable supernatural attributes of the SB fail to give the SB any advantage—they may even prove to be a disadvantage to the SB—and so it would be impossible for a human to determine that the opponent actually is an SB. In those cases, the SB remains hidden from discovery. As Brams puts it, this provides rational support for agnosticism in those who reject outright belief but who, like the French mathematician Blaise Pascal (1623–1662), are also reluctant to firmly declare their disbelief.

'Playing games with God' is not a modern indulgence; as Brams observes, Pascal used such a game-theoretic approach (although he didn't call it that) more than three centuries ago, in his famous analysis on the rational basis for believing in God. Briefly, if God exists and you believe, then you *gain* infinite bliss for all time to come (presumably in Heaven). If you believe and He doesn't exist then you don't lose (or gain) much (if anything). If God exists and you don't believe you *lose* infinite bliss, while if God doesn't exist and you don't believe you don't lose (or gain) much (if anything). The rational choice is obvious—be a believer. The one thing an agnostic knows for sure: you never know, so better safe than sorry! This line of reasoning does seem to have a fatal flaw, however, as surely an omnipotent God would be aware of the spiritual emptiness of Pascal's proclaimed belief.

To believe in God requires that one accept reports of miracles (the virgin birth of Jesus, and his rising from the dead, for example[4]), a step that is just too big for many to take and so they remain 'skeptical.' The patron saint of skeptics, the Scot David Hume (1711–1776), devoted a section of his 1748 masterpiece *An Enquiry Concerning Human Understanding* to how a rational person should react to the claim a miracle has occurred. *By definition*, proclaimed Hume, a miracle violates scientific law and, since scientific laws are rooted in "firm and unalterable experience," any violation of one or more scientific laws immediately provides a refutation of the reported miracle.

The historical motivation for Hume's views was that of what he took to be non-rational arguments for believing in God. As one writer put it,[5] Hume was "an exposer of bad arguments in rational theology." For Hume, second-hand (or even more remote) tales of the return of a man from the dead—the claim that literally defines Christianity ever since Jesus' execution on the Cross—are suspect. As Professor Heath wrote, "Hume . . . makes no attempt to deny the supposed facts; he simply argues that they are consistent with other explanations . . . of a less ambitious kind. There is no right to attribute to the causes of such phenomena abilities more extensive than are needed to produce the observed effects." This is, of course, a view that long pre-dates Hume, a view that goes back to the well-known philosophical concept called *Occam's razor*. It can be found, for example, at least in spirit (no pun intended) in the *Summa Theologica* of Thomas Aquinas (1225–1274). Aquinas actually used the principle of 'make no assumptions beyond what is required' to 'prove' the non-existence of God—and then rebutted his own 'proof!—a theological irony that no doubt didn't escape Hume's notice when he arrived on the scene a few centuries later.

From the very start Hume has had his critics. Many have argued that he wouldn't have been convinced of God's existence by *anything*. One of them, C. S. Lewis, expressed his frustration with Hume in the following amusing way in his 1986 book *The Grand Miracle*: "If the end of the world appeared in all the literal trappings of the Apocalypse; if the modern materialist [Lewis' word for a skeptic] saw with his own eyes the heavens rolled up and the great white throne appearing, if he had the sensation of being himself hurled into the Lake of Fire, he would continue forever, in the lake itself, to regard his experience as an illusion and to find the explanation of it in psychoanalysis, or cerebral pathology."

---

[4] See 1 *Corinthians* 15, in which Paul the Apostle declares the resurrection of Jesus to be *the* basis for the truth of Christianity.
[5] Peter Heath, "The Incredulous Hume," *American Philosophical Quarterly*, April 1976, pp. 159–163. Heath was a professor of philosophy at the University of Virginia.

Lewis is a particularly interesting writer for us—he'll appear again later when we get to religious time travel—because not only was he a witty and persuasive writer on theological matters, he also wrote classic, masterful fantasy (the 1950 *The Lion, the Witch, and the Wardrobe*) and science fiction (the 1938 *Out of the Silent Planet*). He was great pals with J. R. R. Tolkien (*Lord of the Rings*), and clearly had a most inventive imagination; his unfinished work *The Dark Tower* is one of the spookiest pieces of fiction I have ever read. And since in this book I care not a bit if you're a skeptic or a believer—only that you can imagine without dogmatic constraint—then we'll embrace Lewis just as enthusiastically as we do such writers as Isaac Asimov (1920–1992) who more than once declared his belief that there is nothing beyond the grave, and Carl Sagan (1934–1996), an agnostic who included a balanced, energetic debate between a skeptical scientist and a religious man on the existence or not of God in his 1985 novel *Contact*. (I'll return to Sagan's fictional debate later in this chapter.)

1.3 God, Fantasy, and Science Fiction

When you see the words *what if* that I used in Author's Note One, combined with *physics and math*, I think the next words that almost certainly popped into your mind were *science fiction* or, if you prefer, although it seems just a bit pompous to me, *metaphysical speculation*. And that's where the subtitle of this book comes from, as it is in the genre of SF (an abbreviation I'll often use from now on for 'science fiction') that we find, among all the various literary forms, the most complete unshackling of constraints on imagination.

Well, maybe I should back-off that assertion just a bit and say the *second* most complete unshackling. *Good* SF does require that a writer not completely and utterly ignore known science. It is often said that in a science fiction story you are allowed, *at most*, one violation of known science (using the well-known 'willing suspension of disbelief'); if a story has more than one violation then it may still be a good story, but it isn't a science fiction story but rather a *fantasy* story. For example, an author can imagine a perpetual motion machine, *or* a time machine, but cannot have *both* gadgets in the same story and still be writing SF. A slight variation on this 'rule' is that it applies only to the 'typical' SF author. Winners of either the annual Nebula or Hugo writing awards (the Nobel prizes of SF) are allowed two violations, while Grandmasters (authors in the class of Isaac Asimov, Robert Heinlein (1907–1988) or A. C. Clarke (1917–2008), for example) are allowed three. In any case, the appearance of demons, vampires, werewolves, ghouls, dragons, angels, magic, ghosts, zombies, fairies, God and/or the Devil (usually bargaining for a human soul) are also dead giveaways that you're reading a fantasy story.

Now these last words on fantasy are not to be interpreted to mean you are therefore reading a poor story. The Devil, for example, makes a valiant but failed attempt at locking-up the soul of a math professor in Arthur Porges' hilariously funny "The Devil and Simon Flagg," which originally appeared in the August 1954 issue of the most literate of the pulp magazines, *The Magazine of Fantasy and Science Fiction*. And from the same magazine (November 1958), in the story "Or the Grasses Grow" by Avram Davidson (1923–1993), we learn the strange fate of crooks who attempt to steal the land of Indians and so run afoul of a vengeful spirit world. Still, with only a few exceptions, all of the fictional tales discussed in this book are SF, not fantasy. (We will, however, despite my earlier words about him, run into the Devil and his minions again in this book.) For examples of each story type in a 'religious' context, see Appendix 2 (fantasy) and Appendix 3 (SF).

Fantasy and science fiction are occasionally dismissed by serious students of theology who make the curious argument that, since such stories are 'made-up' tales—and so must be telling 'lies'—then those tales are implying either the inadequacy or the outright falsehood of the Biblical tales. This argument (a wrong-headed, indeed ludicrous one) has been specifically addressed by at least three scholars of both genres, two of them academics and the third a writer. The writer, Robert Silverberg (born 1935), wrote the following in an insightful (as well as often hilarious) 1971 essay[6]:

"The problem that arises when you try to regard science fiction as adult literature is that it's doubly removed from our 'real' concerns. Ordinary mainstream fiction, your Faulkner and Dostoevsky and Hemingway, is by definition made-up stuff—the first remove. But at least it derives directly from experience, from contemplation of the empirical world of tangible daily phenomena. ... What about science fiction, though, dealing with unreal situations set in places that do not exist and in eras that have not yet occurred? Can we take the adventures of Captain Zap in the eightieth century as a blueprint for self-discovery? Can we accept the collision of stellar federations in the Andromeda Nebula as an interpretation of the relationship of the United States and the Soviet Union circa 1950? I suppose we can ... But it's much easier to hang in there with Captain Zap on his own level, for the sheer gaudy fun of it. And that's kiddie stuff. Therefore we have two possible evaluations of science fiction: (1) That it is simple-minded escape literature, lacking relevance to daily life and useful only as self-contained diversion; (2) That its value is subtle and elusive, accessible only to those capable and willing to penetrate the experiential substructure concealed by those broad metaphors of galactic empires and supernormal powers."

---

[6] "The Science Fiction Hall of Fame," reprinted in Silverberg's *Beyond the Safe Zone*, Donald I Fine, Inc. 1986.

In a more recent essay, the first academic (Northwestern University professor of media ethics Loren Ghiglione) confessed that "I long dismissed science fiction as fairy-tale foolishness banged out by hacks for barely literate adolescents. Such fiction was aimed at pimply teenage boys who purchased or purloined their sci-fi paperbacks from the bus-stationed racks next to displays of romance novels and the hardcore men's magazines in brown wrappers."[7] It is nothing less than astonishing how many of those "barely literate adolescents" are among today's professional scientists and mathematicians!

As for the second academic, she writes[8] of the secular reputation of SF that "Contemporary science fiction is often negative towards religion. . . However, it is an ideal form to deal with religious themes because it is, by nature, more interested in ideas such as the future of mankind or the ethical implications of science than many other genres. It is thus a natural type of literature to speculate about religion on other planets or in the future."

Let me give you two examples of what the second academic may have had in mind. First, playing on the German philosopher Friedrich Nietzsche's (1844–1900) theme that 'God is dead,' is the dark 1967 story "Evensong" by Lester del Rey (1915–1993). There we find a still powerful entity on the run from those it calls the Usurpers, creatures that are relentlessly hunting him from galaxy to galaxy. As we read we learn of the entity's growing despair as the hopelessness of its fate becomes ever more apparent until, finally, it is trapped by one of the hunters at the very place that the hunt began long ago. As the entity seeks to hide in the dense undergrowth of a garden, it hears the command "Come forth! This earth is a holy place, and you cannot remain upon it. Our judgment is done and a place is prepared for you. Come forth and let me take you there!"

And then we learn the truth, as the story concludes:

"'But—' Words were useless, but the bitterness inside him forced the words from him. 'But why? I am God!'"

"For a moment, something akin to sadness and pity was in the eyes of the Usurper. Then it passed as the answer came. 'I know. But I am Man. Come!'"

"He bowed at last, silently, and followed slowly as the yellow sun sank behind the walls of the garden. And the evening and the morning were the eighth day."

For a second example of other-worldly fantasy that enthusiastically *embraces* Christian theology, consider the well-known saga of Superman, a story known literally around the world that has had a tremendous cultural impact.

---

[7] See "Does Science Fiction—Yes, Science Fiction—Suggest Futures for News?" *Daedalus* Spring 2010, pp. 138–150, for why Professor Ghiglione came to change his mind.
[8] Martha C. Sammons, *'A Better Country': the worlds of religious fantasy and science fiction*, Greenwood Press 1988, p. 127. Sammons was a professor of English at Wright State University.

Our hero's story begins just after his birth on the doomed planet Krypton, as the son of that world's master scientist Jor-El. To save his son from the planet's predicted imminent extinction by explosion, Jor-El sends him by rocket to Earth with the Biblical-sounding good-by of "All that I have I bequeath you, my son. You'll carry me inside you all the days of your life. You will see my life through yours, and yours through mine. The son becomes the father and the father the son." After the rocket arrives on earth the baby is found by a kindly, childless couple and raised as their own. It is a wonderful childhood.

Once he reaches age 18, however, Super*boy* feels an irresistible urge to seek out the Fortress of Solitude in the remote Arctic where he undergoes further training from holographic images of the long-dead Jor-El (a process nicely portrayed in the 1979 movie). The purpose of this training is explained by Jor-El, again in Biblical-sounding words: "They only need the light to show them the way. For this reason, and this reason only, I have sent you, my only son." When he is 30, the now Super*man* returns to the world to begin his new life as 'savior' of humankind. The parallels with the story of Jesus sent to Earth by God are simply too obvious to miss.

Fantasy generally has an attractive, 'romantic' sense to it, but SF is often thought of as escapist, hyper-speculative, unrealistic, gadget-littered writing inhabited by zero-dimensional robots and mutant monsters, one-dimensional humans, and fourth (or higher)-dimensional aliens, all continuously explaining to each other—so readers will know, too—how all their fantastic, futuristic gadgets work. To understand why this particular story-telling goof is the red-flag signature of amateurish writing, just ask yourself how many times *you* have engaged in lengthy discussions with friends about the details of how a jet airplane, a digital video recorder, the telephone network, a refrigerator, a machine gun, the remote control for a garage door opener, or an Xbox360 works? C. S. Lewis called tales like this "Engineers' Stories." In an elaboration of Lewis' characterization, SF writer James Blish (1921–1975) complained that "engineers-turned-writers" often simply cannot resist trying to show their characters are witty sophisticates by having them engage in long, painful, sophomoric back-and-forth banter while leaning over drafting boards design-ing spaceships, or while discussing technical spec-sheets of some fantastic gadget.[9]

Grotesque, salivating, bug-eyed monsters (BEMs in SF lingo) terrorizing screaming (yet always hauntingly beautiful), half-naked Earth women seemed to be uncommonly present in the SF stories of the 1920s through the 1950s or so. This particular imagery was a real favorite for SF magazine cover artists

---

[9] See Lewis' essay "On Science Fiction" in his *Of Other Worlds*, Harcourt 1975, and Blish's essay "Cathedrals in Space," reprinted in *Turning Points* (Damon Knight, editor), Harper and Row 1977.

(who were, without exception, male), grumbled highbrow critics of the genre. Exploration of a philosophical and spiritual nature was imagined by these unhappy analysts to have no place in SF. It was writing appealing mostly to immature teenage boys, so said those demanding readers. As one critic put it, SF is "a genre noted for stereotyping men as Messianic heroes able to conquer everything before them—the moon, the planets of this solar system, the entire galaxy."[10] In much of early SF such a harsh characterization was more or less deserved.[11] There are, however, numerous counter-examples of brilliant, modern SF stories tackling profound issues that actually are *best* set in a science-fictional 'world.' Representative of such tales are those involving speculations of a religious nature.

To support that claim, later in this book I'll discuss many of those stories including (just to excite your curiosity right now, without delay):

1. "The Star" (Arthur C. Clarke): what if it was discovered that the Star of Bethlehem was a supernova that destroyed the planet of a vibrant, advanced civilization?
2. "The Rescuer" (Arthur Porges): could a time traveler armed with a modern, high-power gun 'save' Jesus from the Crucifixion?
3. *Behold the Man* (Michael Moorcock): what if a time traveler discovered that the historical Jesus was unlike the Biblical Jesus, and so assumed the recorded role for himself?
4. "Reason" (Isaac Asimov): what if an intelligent robot decided that it was simply too complicated to have been created by humans, and therefore there must be a 'higher being' beyond humans (and so assumed the role of that being's Prophet)?
5. "Let's Go to Golgotha!" (Garry Kilworth): what if the Crucifixion was simply a tourist attraction for time travelers from all across time?
6. *A Case of Conscience* (James Blish): imagine that a Jesuit biologist, a member of a mission to an alien planet, finds that the natives are intelligent, supremely rational beings who have no concept of God. There is no evil, no religion, and no original sin—is this utopian world the work not of God but of Satan, one created to show humans that a belief in God is not necessary for a happy and good life?
7. *"Angel of the Sixth Circle"* (Gregg Keizer): what might be the implications of a time traveling hit-man, in the employ of a new religious movement, who

---

[10] Beverly Friend, "Virgin Territory: the bonds and boundaries of women in science fiction" in *Many Futures, Many Worlds: theme and form in science fiction* (Thomas D. Clareson, editor), The Kent State University Press 1977, pp. 140–163.

[11] But not always. For example, H. G. Wells' *The Time Machine*, first published in 1895 in serial form, is deservedly recognized as a literary classic and it has never been out-of-print.

kills Catholic priests in the past to improve the situation of the movement in the present?

8. "The Word to Space" (Winston P. Sanders[12]): Suppose that the first contact with an alien culture is by radio, with a planet 25 light-years distant that is ruled by a fanatical religious theocracy that attempts to subvert all of Earth's religions by inundating us with endless religious broadcasts? How, in particular, would the Vatican respond?

And where else but in a science fiction story—as in the 1951 tale "The Quest for Saint Aquin" by Anthony Boucher (1911–1968)—could you have the following exchange between Thomas (an emissary of a Pope in the far future) and an intelligent *robot?*:

"To believe in God. Blah. .... I have a perfectly constructed logical mind that cannot commit such errors."

"I have a friend," Thomas smiled, "who is infallible, too. But only on occasions and then only because God is with him."

"No human being is infallible."

"Then imperfection," asked Thomas, suddenly feeling a little of the spirit of the aged Jesuit who had taught him philosophy, "has been able to create perfection?"

"Do not quibble," said the [robot]. "That is no more absurd than your own belief that God who is perfection created man who is imperfection."

As the science fiction writer and anthologist George Zebrowski has written[13]: "Science fiction as the art of the hypothetical has been in a unique position to speculate freely about religious concepts. A story is not bound by the strictures of assertion or argument, and a writer need not believe in the conditional circumstances of the possibility or impossibility depicted in his work. .... a science fiction writer can try a concept on for size, spinning out its imaginative content in any direction. The creativity of science fiction asserts only the autonomy of the disciplined imagination, the pursuit of concepts and constructs and their expression in pleasing narrative forms, often purely for their aesthetic and intellectual beauty."

In other words, these are tales that will make you *think.* Stanislaw Lem wrote of this signature feature of SF in one of his many insightful essays[14]: "As in life we can solve real problems with the help of images of non-existent beings [for an example of what I think Lem had in mind, astronomers solve for

---

[12] A penname for a writer better known under his real name, Poul Anderson.

[13] From the Introduction to the anthology *Strange Gods* (Roger Elwood, editor), Pocket Books 1974.

[14] Lem's "On the Structural Analysis of Science Fiction," *Science-Fiction Studies*, Spring 1973, pp. 26–33.

the orbits of the planets by thinking of those spatially *extended* objects as idealized *point* masses], so in literature we can signal the existence of real problems with the help of prima facie impossible occurrences of objects. Even when the happenings it describes are totally impossible, a science fiction work may still point out meaningful, indeed rational problems."

I think you'll discover on the following pages of this book just how many SF writers have thoughtfully treated a wide spectrum of religious themes, and that the view of the time traveler in the 1986 novel *Moscow 2042* by Vladimir Voinovich (born 1932) is *not* necessarily true: "Science fiction ... is not literature, but tomfoolery like the electronic games that induce mass idiocy." There certainly is a lot of nonsense in science fiction, but there is a lot of nonsense in *every* genre. The point, not to state too bluntly the obvious, is to be discriminating.

1.4 God and Science

I mentioned earlier that one of the attributes Steven Brams assumed for his 'supreme being' was omnipotence. As an excellent example of how the mind of a technically trained SF writer works, 'what if' God did *not* have that attribute? That's the question Arthur C. Clarke asked himself in a 1972 essay, in which he put forth what he called an "astrotheological paradox."[15] After observing that the speed of light is the absolute limiting propagation speed of information and energy, Clarke writes

> "If God obeys the laws He apparently established, at any given time He can have control over only an infinitesimal fraction of the Universe. All hell might (literally?) be breaking loose ten light-years away ... and the bad news would take at least ten years to reach Him. And then it would be another ten years, at least, before He could get there to do anything about it."

Clarke isn't ignorant of an obvious rebuttal to this: "Nonsense, God is already 'everywhere'." This is sometimes called 'divine omnipresence,' an attribute that seems to be claimed in the Old Testament (*Jeremiah* 23:23–24) where we read "*Am* I a God near at hand, saith the LORD, and not a God afar off? Can any hide himself in secret places that I shall not see him? saith the LORD. Do I not fill heaven and Earth? saith the LORD." Despite this endorsement in Scripture, there are those who nevertheless view omnipresence with suspicion.

---

[15] "God and Einstein," in Clarke's *Report On Planet Three and Other Speculations*, Harper & Row 1972, pp. 115–116.

In one analysis,[16] for example, we read that the "attribute of omnipresence . . . has always been something of an embarrassment to classical supernaturalism . . . The presence of God is indeed something unique . . . He is the only entity who can be said to be everywhere: *but what does this mean?* In the hands of supernaturalist theologians the attribute of omnipresence does seem to die 'the death of a thousand qualifications.'"

Now, as I stated earlier, this is a 'what-if' book and so—*just suppose*—that God's awareness of events *is* limited by Einstein's relativity theory. With that supposition then, Clarke ends his essay with these chilling words: "He's coming just as quickly as He can, but there's nothing that even He can do about that maddening 186,000 miles a second. It's anybody's guess whether He'll be here in time."

In the 1950s Clarke wrote two short 'theological' stories which have since been anthologized numerous times ("The Nine Billion Names of God" and one I mentioned earlier, "The Star"), both of which are discussed later. He once remarked that after the great English biologist (and self-proclaimed atheist) J. B. S. Haldane (1892–1964) had read "The Nine Billion . . ." he wrote to Clarke to say "You are the only person to say anything original about religion for the last two thousand years." (Haldane was the model in C. S. Lewis' *Out of the Silent Planet* for the villain/genius physicist Weston—"Has Einstein on toast and drinks a pint of Schrödinger's blood for breakfast"— whose only 'religion' was that of science). I think Haldane's praise just a bit over the top, but in fact Clarke's 1972 essay might fairly be counted as a non-trivial addition to modern theological speculation: it offers us, for example, a *physical* reason, different from the *mathematical* one due to Brams, for understanding the presence of evil in a universe created by a benevolent God who is also constrained by His own laws.

I should mention that earlier SF stories did anticipate Clarke's essay, although none made his points with the same explicit force. For example, in a 1942 story that rewrites *Genesis* (I'll discuss this tale more completely later), "The Cunning of the Beast" by Nelson Bond (1908–2006), God is stated in passing to move "with the speed of light." And in the story "Shall the Dust Praise Thee?" by Damon Knight (1922–2002), God arrives on Earth for Judgment Day, only to find the planet utterly destroyed by nuclear war and these words on the wall of an underground bunker: "WE WERE HERE. WHERE WERE YOU?" (So here we have a non-omniscient God showing up late for the Apocalypse!) When Knight's story first appeared in a 1967 anthology he included the following provocative Afterword: "This story was

---

[16] Rem B. Edwards, *Reason and Religion*, Harcourt Brace Jovanovich 1972, p. 175. Edwards was a professor of philosophy at the University of Tennessee.

written some years ago, and all I remember about it is that my then agent returned it with loathing, and told me I might possibly sell it to the *Atheist Journal* in Moscow, but nowhere else. The question asked in the story is a frivolous one to me, because I do not believe in Jehovah, who strikes me as a most improbable person; *but it seems to me that, for someone who does believe, it is an important question* [my emphasis]."

Lest I give you the impression that a scientist writing religious SF has some sort of 'imagination advantage' over non-scientists, let me give you an example of what I think an astonishing *goof* from none other than the well-known astronomer Carl Sagan. I mentioned earlier that in his novel *Contact* there is a quite interesting debate between a skeptic and a believer and, as part of this encounter, Sagan has his skeptic challenge the believer with a question that has bothered many—even theologians—for centuries: why doesn't the Bible contain any clear, unambiguous statements that would absolutely convince *anyone, at any time in history,* that the words truly come from a supernatural being?

In just a moment I'll tell you where Sagan goes with this in *Contact*, but first here is what Professor Heath suggests: "What is wanted, evidently, is . . . communication on a scale which only omnipotence could account for. Crude as it may be, the following flight of fancy will perhaps bring the matter to a head: if the stars and galaxies were to shift overnight in the firmament, rearranging themselves so as to spell out, in various languages, such slogans as **I AM THAT I AM**, or **GOD IS LOVE**—well, the fastidious might consider that it was all very vulgar, but would anyone lose much time in admitting that this settled the matter . . .? Confronted with such a demonstration, the hard-line Humean could continue, of course, to argue that, for all its colossal scale, the performance is still finite, though immense, and so cannot be evidence of more than the finite, though immense, power that is needed to achieve it. But this now seems a cavil, designed only to prove that even omnipotence is powerless against the extremer forms of skeptical intransigence. . . . If celestial inscriptions of this kind had first appeared, say, at the time of the Crucifixion, and were periodically altered in accordance with terrestrial circumstances, it would long since have become entirely natural to treat them as a system of messages emanating from a supreme being, and as the clearest evidence imaginable of the authenticity of a particular revelation."

When Sagan's skeptic is challenged by the believer to give examples of what sort of statements could have been put in the Bible to convince anybody, the skeptic replies with a version of just what Professor Heath suggests (without moving the stars about!). The following examples would, as Sagan writes, "leave a record for future generations" that would make God's "existence unmistakable": "The Sun is a star," or "A body in motion tends to remain in motion," or "There are no privileged frames of reference," or "Thou shalt not travel faster than light." As Sagan has his skeptic declare, all of these statements, in addition to sounding 'cryptically Biblical' to the people of the times of Moses or Jesus, are true statements that nobody could "possibly have known three thousand years ago." (I am sure any reader of this book could cook-up more such statements, too. Here's one of my own: "The oceans are the union of two, one without which no man or beast can live, and the other is two-fold and the lightest of all." That would have meant nothing to the followers of Jesus, but a lot to any high school chemistry student who has just learned what can result when hydrogen and oxygen get together in the formula H_2O.) The skeptic's views in the novel were Sagan's personal views, as he repeated them, word-for-word, in his Gifford Lectures on Natural Theology that he gave in 1985 at Glasgow University.[17]

This is excellent skeptical stuff, I think, but Sagan's admirable approach is not to preach to the reader, but rather to be even-handed, and so his believer gives as good as he gets. At the end of the debate the skeptic admits to a friend "I don't think I did much to convert him. But I'll tell you, he almost converted me." So far, so good. But then Sagan, at the end of his novel, goes astray by having his skeptic make the computer-aided 'discovery' of a secret message tucked away in the infinite digits of the transcendental number pi, a message Sagan calls "The Artist's Signature." The numerical value of pi is intimately connected with the geometry of a circle (it's the ratio of the circumference to the diameter, of course), and the big discovery is that there is, somewhere in pi's digits, a long, unbroken string of just 0s and 1s such that, if printed out as a square array, forms the picture of (big drum roll) . . . holy cow, *a circle*! Any mathematician reading that must have ground his/her teeth with the utter banality of such a so-called 'message.'

Sagan did include a bit of discussion of all the obvious objections to the sensibility of the 'big discovery'—but then waved them away with what I think some quasi-mathematical irrelevances. Here's what I mean by that. The digits of pi are believed to be what mathematicians call 'uniformly random,' which simply means that if you look at very long strings of digits you'll see 0 through 9 each randomly occur about 10 % of the time. Therefore, if you look long enough you'll find *all* possible sequences of those digits, including *all* possible sequences of just 0s and 1s. Sagan has one of his characters point this out, and then has the discoverer of 'God's message' brush that aside with the comment 'Yes, of course, but it will be a very low probability occurrence if it happens just by chance.' That's simply mathematical nonsense—*all* possible sequences *will occur with probability 1*. The pictures that these strings of 0s and 1s would

---

[17] The lectures can be found in Sagan's *The Varieties of Scientific Experience: a personal view of the search for God*, The Penguin Press 2006.

form will not be just Sagan's circle, but also *all other possible images*, including the Nazi swastika, a message not from God but rather, if from anyone, from Satan. I can only wonder how many people who read *Contact* actually came away thinking there really is a secret message from God in pi. When the book was made into a 1997 film, the screenwriters wisely decided to delete that scene from the script.

There is a wonderful passage in a famous SF novel by Robert Heinlein—his 1961 *Stranger in a Strange Land*—that I think nicely catches my view of the interlocked nature of science (and science fiction) and theology. The words are those of a character who befriends a man born on Mars, a man who then travels to Earth (the 'Strange Land' of the title, which Heinlein apparently took from *Exodus* 2:22): "Man is so built that he cannot imagine his own death. This leads to endless invention of religions. While this conviction by no means proves immortality to be a fact, questions generated by it are over-whelmingly important. The nature of life, how ego hooks into the body, the problem of ego itself and why each ego *seems* to be the center of the universe, the purpose of life, the purpose of the universe—these are paramount questions . . . they can never be trivial. Science hasn't solved them—and who am I to sneer at religions for *trying* . . .?"

1.5 Author's Note Two

In conclusion of this introductory chapter, let me say a little about the origin of the book. The idea for its writing came to my mind slowly, over a period of many years. The question of the historical evidence for the actual existence of Jesus has long fascinated me: did a person of that name *really* exist and do all the things the New Testament describes or, like those of King Arthur and Robin Hood, are the stories in the four Gospels simply ones that we all just *wish* could be true? Much of the story of Jesus is not unique to him; the mythologies of the ancient Greek gods Osiris, Attis, and Dionysus, for example, all include the idea of resurrection and, like Jesus, Dionysus in particular was born of a mortal woman and a supernatural father (Zeus). This, all *centuries* before Jesus.

The Bible is said to be the inspired word of God, but of course the words we read today have passed through the hands of many men, translators, and interpreters over the past 2,000 years, each with their own agenda.[18] For

---

[18] See, for example, Richard Elliott Friedman, *Who Wrote the Bible?*, Harper & Row 1987. Freidman is a distinguished Bible scholar, now on the faculty of the University of Georgia. More provocative reading can be found in Robin Lane Fox, *The Unauthorized Version: truth and fiction in the Bible*, Viking Penguin 1991, and Bart D. Ehrman, *Misquoting Jesus: the story behind who changed the Bible and why*,

those who think the Bible is literally the 'word of God,' remember that we do not have the original documents that form the Bible; all we have are copies of copies . . . of copies, from Aramaic to Greek to . . . to Coptic to Latin to English, with 'adjustments' at every stage. It's a bit like the story of the fellow who claims to have the very axe his grandfather used a hundred years ago: the handle has been replaced only three times and the blade only twice!

An example of what I am getting at here can be found in the second volume of Isaac Asimov's autobiography,[19] in which he writes of an experience he had after writing his well-known *Guide to the Bible*. While on a visit with a friend to Brandeis University near Boston, Massachusetts, to view a collection of old Bibles, Asimov writes

"At one point we were looking at a Jewish Bible published in Spain before the expulsion of the Jews. It was open to the seventh chapter of *Isaiah*, and was in Spanish, except for one word that was in Hebrew and stood out like a sore thumb amid all the rest."

"My friend said to me, 'Why do they have one word in Hebrew?'"

"Having spent some time on that very point in my Bible book, I said 'That's the verse that, in the King James, goes, 'Behold a virgin shall conceive, and bear a son.' The only trouble is that the Hebrew word is *almah*, which does not mean 'virgin' but 'young woman.' If the Jewish publishers were to translate the word correctly they would seem to be denying the divinity of Jesus and they would be in serious trouble with the Inquisition. Rather than do that, or translate incorrectly, they leave that word in Hebrew."[20]

In another place Asimov had a funny illustration of those who don't fully appreciate the evolutionary history of the Bible. There[21] he wrote of one aged parishioner who said, while waving his Bible, "If the King James was good enough for the prophets and apostles, it is good enough for me." This would, I suppose, make some sense in a world with time travel.

A modern SF classic, treating this question of ancient history by imagining what the Catholic Church might evolve into in the far-distant future, is the 1979 short story "The Way of Cross and Dragon" by George R. R. Martin

---

HarperCollins 2005. Ehrman is a professor of religious studies at the University of North Carolina, Chapel Hill, and Fox is a historian at Oxford. Most recent is the quite scholarly (that means it's pretty dense reading!) by Philip Jenkins, *Jesus Wars: How Four Patriarchs, Three Queens, and Two Emperors Decided What Christians Would Believe for the Next 1,500 Years*, HarperCollins 2010. The title says it all. Jenkins is a professor of history and religious studies at Penn State University, and a senior fellow in religious studies at Baylor University.

[19] *In Joy Still Felt*, Doubleday 1980, p. 461.

[20] The verse Asimov is discussing is *Isaiah* 7:14, and you can find additional discussion of it in his *Asimov's Guide to the Bible: the Old and the New Testaments*, Avenel 1981, p. 532.

[21] In *Creations: the quest for origins in story and science* (Issac Asimov, *et al.*, editors), Crown 1983, p. 101.

(born 1948).[22] The tale is set so far in the future that nearly two centuries have passed since the Pope (of the One True Interstellar Catholic Church of Earth and the Thousand Worlds) ruled that non-humans might serve in the clergy.[23] The present Pope is an immense, whale-like alien creature that, smelling like rancid-butter, floats naked (save for a damp Roman collar) in a pool. He has summoned Father Damien, a human Inquisitor of the Order Militant of the Knights of Jesus Christ, to order him to investigate the charge of heresy on the planet Arion, where Judas Iscariot, shockingly, is celebrated as a *saint*.

As Father Damien travels to Arion in his faster-than-light spaceship (named *Truth of Christ*) he reads that world's heretical bible, called *The Way of Cross and Dragon*. It is a mish-mash conglomeration of myths, legends, and the Church's own Bible, relating the life of Judas and his relationship with Jesus. A Keeper of Dragons, Judas was actually (according to *The Way*) the victim of lies spread by Peter. On Arion, however, the 'truth' is at last being preached by a rogue priest who has broken free of the One True Interstellar Catholic Church.

Father Damien is enthralled by the stories in *The Way* but, of course, thinks them ultimately absurd. As he tells a friend (who has also read *The Way*) during the journey to Arion, it's all "an unbelievable tangle of doctrine, apocrypha, mythology, and superstition. Entertaining, yes, certainly. Imaginative, even darling. But ridiculous, don't you think? How can you credit dragons?" In reply, his friend laughs and says "Is that any sillier than water changing into wine, or Christ walking on the waves, or a man living in the belly of a fish?"

Soon after arriving on Arion Father Damien confronts the priest with the words "A more ridiculous creed I have yet to encounter. I suppose you will tell me that you have spoken to God, that he trusted you with this new revelation, so that you might clear the good name, such that it is, of Holy Judas?" He is stunned when the reply is (along with a smile) "Oh, no, no, I made it all up."

When asked to explain *why*, the priest replies that it is simply to make the people of Arion happy. The truth would be far too hard for most to accept, and the truth is that there "is no afterlife, no God. [Those who know the truth] see the universe as it *is* . . . and these naked truths are cruel ones. We who

---

[22] Martin is best-known today, to viewers of television's HBO, as the creator of that channel's extraordinarily popular fantasy series *The Game of Thrones*.

[23] This idea, of aliens not only being proselytized by human missionaries but actually becoming members of the missionary faith, was presented in an hilarious 1974 story by William Tenn (the pen-name of Philip Klass (1920–2010)), "On Venus, Have We Got a Rabbi." There we learn of the very non-human Bulbas, from the fourth planet of the star Rigel, who with the help of a human rabbi win the legal right to be recognized as Jews.

believe in life, and treasure it, will die. Afterward there will be nothing, eternal emptiness, blackness, nonexistence. In our living there has been no purpose, no poetry, no meaning. Nor do our deaths possess these qualities. When we are gone, the universe will not long remember us, and shortly it will be as if we had never lived at all. Our worlds and our universe will not long outlive us. Ultimately, entropy will consume all, and our puny efforts cannot stay that awful end. It will be gone. It has never been. It has never mattered. The universe itself is doomed, transient, uncaring." Father Damien professes to be appalled at that but, by the end of the story and even as he continues to serve his bloated, alien Pope in rooting out heresy from one end of the galaxy to the other, his *faith* has been destroyed.

So, to repeat my question: the Biblical stories are undeniably ones of powerful moral teaching but, ultimately, are they just *stories* to make us happy? It is, after all, only by the Gospel stories (which don't always agree), and the account given by the historian Flavius Josephus (c.37–100 AD) in his *The Antiquities of the Jews*, that we have any record at all for a person called 'Jesus of Nazareth.' And even then it is only for the time of his ministry that anything of detail is told. Still, Josephus lived in Galilee and knew people who were alive when Jesus was said to be active, and so it *seems* as though the historical truth of Jesus is on reasonably strong ground. But the question still tantalizes.

But it can be a risky question to ask. For how an analysis of a story of Jesus that deviates from traditional Church teachings on his divine origin (even if sympathetic) can provoke the outrage Asimov suggests, I remind you of Hugh Schonfield's *The Passover Plot*. When the book originally appeared in 1966 it was a sensation, offending as many as it did whose imaginations it captured. In it we are told that Jesus did not die on the Cross, but rather craftily *pre-arranged* for all the post-Crucifixion events that we read about in the Bible. The idea that Jesus didn't die on the Cross actually long predates Schonfield; you can find it, for example, in the *non*-SF 1929 novella *The Man Who Died* by D. H. Lawrence. Lawrence's work is actually even more provocative than is Schonfield's, as he has Jesus surviving his ordeal by accident and then, realizing what has happened, Jesus renounces any further attempt at preaching the word of God and sets forth into the world to enjoy all the earthly pleasures he had missed during his ministry (including having a child).[24]

The same idea was used in the 2007 story "Friends in High Places" by Jack McDevitt (born 1935), which opens with Jesus waiting in the Garden of

---

[24] In a case of 'fact following fiction,' in September 2012 Karen King, an historian of early Christianity at the Harvard Divinity School, announced the discovery of a fourth-century Egyptian papyrus that refers to Jesus as being married. That announcement caused, as you'd imagine, not just a little turmoil at the Vatican.

Gethsemane for the mob to take him. Jesus does *not* want to die, as we learn from his thoughts:

> "It sends the wrong message [Lord]. It will be a hard sell, persuading people You love them when you let this happen to me."
>   and
>   "Why? Why must we do it this way? We create a faith whose governing symbol will be an instrument of torture. They will wear it around their necks, put it atop their temples. Is this what we really want?"

In this story, too, Jesus escapes (to become a librarian in Egypt!), and as he begins his journey to a new life he thinks "how much better it was than a cross." What has happened is that God, apparently in answer to Jesus' concerns about the Crucifixion, has *changed the past*. We suspect something like this has happened because the Greeks, not the Romans, are in power, and then the really big clue comes when Jesus learns that decades earlier Mark Antony had *won* the naval battle of Actium. (Whether or not changing the past makes any 'sense' is an issue I'll address in a later chapter!)

There are some who think the question 'Was Jesus a real person?' is meaningless, as silly to ask as might be 'Did Shakespeare actually write *Julius Caesar*?' (or was it, as the old joke goes, somebody else with the same name?), or as might be the question 'Did Homer really write the *Illiad*?' We have those great works in front of us today, so goes the argument, so let's just read them and who cares who actually wrote the words hundreds or thousands of years ago. I think the analogy of the mystery of authorship and of the reality of Jesus to be a grossly false one. The books and plays of Homer and Shakespeare *do* exist, yes, and in the end it is the greatness of their ideas that is what really matters.

The existence or not of Jesus is not so smoothly dismissed, however. If he didn't actually live then what we are left with is nothing less than a stupendous fraud 2,000 years in the making, in the name of which literally millions have died. The question of Jesus' historical reality *does* matter and, as you'll see in the pages of this book, science fiction writers have eagerly tackled that question and more, sometimes in ways that may be shocking. In the opening to an anthology of religious science fiction stories, for example, the editor wrote "I am perfectly certain in my own mind that God is alive. I am less certain that He's well. I think, in fact, that He may be fighting for His life in the pages of this book."[25]

As an academic electrical engineer who has bought (and written, too) a lot of math and physics books, I've also amassed a fair number of texts in my personal library on Biblical archaeology. As you might expect from that, I

---

[25] See the Introductory essay by Alan Ryan in *Perpetual Light*, Warner 1982.

am a great fan of the *Indiana Jones* movies—or at least of the first (and best) one, the 1981 *Raiders of the Lost Ark*—and I think it was Indy's quest for the Ark that really planted the seed for this book. (But see Appendix 3 for how my interest in such matters predates the film.) The Ark in the movie was not Noah's Ark, but rather the Ark of the Covenant (or Ark of the Law) which was built to transport the stone tablets bearing the Ten Commandments, as received in the thirteenth century BC by Moses from God on Mt. Sinai. The Ark, built by Moses according to detailed instructions from God (*Exodus* 25), is described in various Jewish legends as being surrounded by sparks and so was seemingly electrical in nature. Further, to directly touch the Ark itself (it was carried on poles), for any reason, was to be immediately struck down, as was Uzzah in the tenth century BC (*Samuel* 6:6–7); perhaps a death by divine electrocution? In *Exodus* 25:22 the Lord tells Moses He will speak to him from the Ark, and in the movie the electrical nature of the Ark is implied when the central villain dramatically tells Indy "It's a transmitter. It's a radio for speaking to God!" The Ark eventually ended-up in the Temple of Solomon in Jerusalem, where it remained until the destruction of that city by the Babylonians in the sixth century BC. At that point the Ark disappears from both the Bible and history—until more than twenty-five centuries later when the Nazis,[26] and then Dr. Jones, take-up the hunt.

One SF explanation for the disappearance of the Ark is given in the 1954 story "For I Am a Jealous People" by Lester del Rey. There we read that God, furious with an unrepentant world, has broken the ancient covenant with man and abandoned him, and has taken an alien race of reptiles to be his new chosen 'people.' An alien priest (who speaks perfect English[27]) tells a human, "The Lord Almighty commanded us to go down to Earth where abominations existed and to leave no living creature under your sun." To aid the aliens as they invade Earth by this divine decree, He has given them the Ark to carry into combat.

The final, big push for me to go forward with this project came when I read Ari Goldman's terrific 1991 book, *The Search for God at Harvard*. Goldman, then the religion reporter at *The New York Times* (he is now a professor of journalism at Columbia University), received a sabbatical leave from the paper to spend a year (1985) at the Harvard Divinity School. The vivid description

---

[26] The movie's depiction of Nazi obsession with the supernatural was not just made-up Hollywood make-believe nonsense. See, for example, Bill Yenne's *Hitler's Master of the Dark Arts: Himmler's Black Knights and the Occult Origins of the SS*, Zenith Press 2010.

[27] The problems of communication between humans and SF aliens have received a scholarly (and highly entertaining, too) treatment by Walter E. Meyers, a professor of English (now emeritus) at North Carolina State University in his book *Aliens and Linguists: Language Study and Science Fiction*, The University of Georgia Press 1980.

of his experiences that year sharply illustrated for me the vast diversity of thought among theologians, and prompted me to take seriously many of the issues that I then realized puzzled not just me, but the 'professionals,' too. The science fiction component of this book is a continuation of the themes of two of my earlier books.[28] To understand how two such *seemingly* distinctly different patterns of thought—SF and religion—can be connected, an argument (that I really like) was nicely made by Robert Silverberg in the Introduction to his short-story collection *Beyond the Safe Zone* (see note 6 again):

> "When the world turns incomprehensible, it makes sense to look for answers from some other world. In former times it was sufficient to look no further than the Church: God was there, emanating love and security, offering the hope of passing onward from this vale of tears to the true life beyond. One of the difficulties of [modern] life is that most of us have lost the option of using religious faith as a consolation. It may be that science fiction has evolved into a sort of substitute: a body of texts of an examination of absolute values and the hypothetical construction of alternative modes of living."

Now, at the risk of being repetitious, let me end this first chapter by restating my opening words in an alternative way, just to be absolutely clear on what my intentions are with this book. I am not a believer, but I certainly do hope that as you read you won't think I'm being obnoxious about it.[29] I am not going to be the bomb thrower that Oxford emeritus fellow and former professor Richard Dawkins is, a man described on the dust jacket of his own 2006 book *The God Delusion* (Houghton Mifflin) as "the world's most prominent atheist." I do, in fact, personally agree with just about everything Dawkins argues in his book, but the rationality for believing (or not) in a personal, supernatural God is *not* what this book is about. It's about how SF has treated religious issues. *And that's all.*

SF writers have a thoroughly imaginative mind-set (if they don't they don't survive as writers for long!), and so many of their religious tales snuggle-up pretty close to what some—usually devout Christians not used to entertaining any challenge at all to what they were taught as kids in Sunday school—

---

[28] *Time Machines: time travel in physics, metaphysics, and science fiction* (2nd edition), Springer-AIP 1999, and *Time Travel: a writer's guide to the real science of plausible time travel*, Writer's Digest Books 1997 (reprinted, with a new Preface, in 2011 by The Johns Hopkins University Press).

[29] As I finish the writing of this first chapter, I think that my take on the matter of God has, in fact, evolved towards the one adopted by the SF writer James Blish: "I believe there might have been a Creator but He never intervenes, does not desire worship and may not even be around any more." Quoted from the brilliant, book-length treatment of Blish by David Ketterer, *Imprisoned in a Tesseract: the life and work of James Blish*, The Kent State University Press 1987, p. 321.

consider impropriety. SF writers are often vocal skeptics. As an afterword to his story "Friends in High Places," Jack McDevitt wrote

> "I've never felt comforted or encouraged by the notion that God would stand by and allow his son to go to the cross. (If that happened to Jesus, what were *my* chances?) Or that he would be willing to watch casually while tidal waves rolled in and killed tens of thousands. Or lethal diseases ravaged whole continents. Or Nazis ran wild and killed millions [McDevitt might have mentioned Stalin's purges at this point, too]. You have to be willing to overlook a lot to accept the idea that a compassionate supernatural force worries about our welfare. But we are capable of doing it. A man misses a plane, the plane goes down, two hundred people die, but the guy left standing in the parking lot starts talking about how God stepped in to save his life. And we buy it [with the explanation, McDevitt might have added, based on the well-known phrase 'God works in mysterious ways']. Never mind the crew and passengers on the flight."[30]

McDevitt is right, that *is* asking a lot, and I personally am in sympathy with his concern. Nonetheless, he still strikes me as less extreme than is Dawkins.

Since I've tried very hard to avoid proselytizing in this book, I am not even going to take the soft position taken by the authors of a recent book-length anthology of essays that collectively examine the issue of the resurrection of Jesus, the central claim of Christianity asserting that he returned from the dead.[31] The essays are quite hostile to attempts to base Christianity on miracles, and yet they remain sympathetic to Christianity. Whether or not *you* accept miracles is not at play in this book, however. If you do, then okay. If you don't, well, that's okay (with me, anyway), too.

A recent book that may, at least superficially, appear similar to this one has as its goals (according to the cover blurb) that of teaching readers "how to think of God" and to help readers "keep their beliefs alive in a world of rapidly changing technology." Now I *did* find that book helpful in my writing (I recommend it as a good read), but while I am sympathetic to its first goal I am going to specifically avoid the second.[32] That's because this book is *not* about *my* beliefs, and I have *no* personal agenda to convert your beliefs to mine. So, *please*, no outraged e-mails in an attempt to convert my beliefs to yours!

---

[30] *A Cross of Centuries: twenty-five imaginative tales about the Christ* (Michael Bishop, editor), Thunder's Mouth Press 2007, pp. 45–46.

[31] *The Empty Tomb: Jesus Beyond the Grave* (Robert M. Price and Jeffery Jay Lowder, editors), Prometheus Books 2005. See also note 4.

[32] Gabriel McKee, *The Gospel According to Science Fiction*, Westminster John Knox Press 2007. The author has a Master of Theological Studies from the Harvard Divinity School, which probably explains the second goal.

Chapter 2
Religious Science Fiction *Before* Science Fiction

2.1 The 'Start' of Modern SF

A close relative of SF is *horror*, another genre that makes great use of the fantastic and a willing suspension of disbelief. Two well-known writers of the modern era who worked in both areas, occasionally in the same story, were H. P. Lovecraft (1890–1937) and Ray Bradbury (1920–2012). Long before the rise of SF to what it has become today, the literary niche occupied by horror stories for the masses was a busy place. Dating back to before the start of the nineteenth century and the invention of the high-speed rotary printing press, the Georgian and Victorian periods in England, in particular, were seemingly populated by endless numbers of people who couldn't get enough of tales involving supernatural entities like ghosts, vampires, the devil, demons, werewolves, and other assorted monsters (a role played later in SF by 'aliens from the stars'). Sex sold well, too, and the Gothic horror novel had a tremendous following, with hack writers often out-selling more recognized authors such as Dickens and Thackeray. Not all such works were prurient, of course, with the 1818 *Frankenstein* by Mary Shelley (1797–1851) considered today to be a classic, as is the much later 1897 vampire horror novel *Dracula* by Bram Stoker (1847–1912).

Scholars date the origin of the horror genre for 'the unwashed and unsophisticated but able to read' with the 1764 appearance of *The Castle of Otranto* by Horace Walpole (1717–1797). It was soon followed by many others of a similar nature, including (in 1796) *The Monk* by Matthew Lewis (1775–1818). The tale of a young monk enthralled with sex and demonology to the point of selling his soul to the Devil, one critic's appraisal of it nicely describes the spirit of the typical Gothic horror tale: "a mass of murder, outrage . . . and indecency."[1] As the decades of the 1800s passed, literally kilo (if not mega) tons of inexpensive

---

[1] Peter Haining, *A Pictorial History of Horror Stories*, Treasure Press 1985, p. 12.

P.J. Nahin, *Holy Sci-Fi!*, Science and Fiction,
DOI 10.1007/978-1-4939-0618-5_2, © Springer Science+Business Media New York 2014

horror story magazines printed on cheap paper were eagerly purchased by readers who simply couldn't get enough of them. Going under the descriptive names of 'penny dreadfuls,' 'penny bloods,' and 'shilling shockers,' names indicating the economic status of their intended audience, those magazines made a lot of money for their publishers.

This success led, as the nineteenth century came to its end, to the creation of a higher caliber magazine, the so-called *slick* (in reference to its better grade of paper) such as *The Strand Magazine* in 1891. It was in *The Strand*, for example, that the wonderful adventures of Sherlock Holmes first appeared. The slick concept soon crossed the Atlantic and appeared in America, and included *Harper's*, *The Century*, and *Scribner's*. All these magazines carried horror fiction (commonly appearing, for example, were demonology and witchcraft in *Harper's*, ghosts and werewolves in *The Century*, and ghosts and explicit torture in *Scribner's*), but only as a *portion* of the contents.

The very first all-fiction pulp was *The Argosy*, begun in 1896 by Frank Munsey (1854–1925), and in 1905 he started *The All-Story Magazine* devoted specifically to adventure tales. (The term *pulp* came from the use of inexpensive wood-pulp—you could *feel* the lumpy wood chips in each ragged, untrimmed page—to make paper that was far too crummy for the use by any publisher of 'words meant to last.' Such paper quickly yellowed, turned brittle, and finally, amid billowing clouds of bits and pieces, entered into eternal oblivion. Think of the paper used in your newspaper before its final contribution to civilization in the bottom of your cat's litter box; pulp was worse.) Both of Munsey's magazines often published stories that, before the term 'science fiction' was coined, went under the general rubric of the 'scientific romance' (as did, for example, the classic SF novels of H. G. Wells, such as *The Time Machine*, *War of the Worlds*, and *The Invisible Man*).

The late 1930s and the 1940s define the period generally thought of as the 'golden age' of pulp magazine science fiction. Before then, however, starting with the April 1926 appearance of the first issue of *Amazing Stories*, there is a precursor decade or so of what Hugo Gernsback (1884–1967), the publisher of *Amazing*, described with the clumsy word "scientifiction." *Amazing Stories* was the first pulp to be devoted totally to science fiction. With its masthead motto of "Extravagant Fiction Today—Cold Fact Tomorrow," and with the illustration on the contents page of each issue showing a muscular Jules Verne (1828–1905) bursting from his grave in the heroic, up-up-and-away pose made famous years later by Superman, there could be no doubt as to what kind of fiction the reader would find under the dramatic, multi-colored cover art.

The stories in *Amazing* were 'read it in the morning, forget it by dinnertime' adventure fiction, the sort of stuff you'd put inside a newspaper if on a crowded train or bus so fellow passengers wouldn't know what a low-grade mind you had. The transient nature of pulp fiction was independent of its literary quality, as the cheap acid-based paper that stories were printed on began to oxidize and literally burn-up as soon as it rolled off the press. In the introductory essay to a 1950 collection of pulp-detective Philip Marlowe stories (*Trouble Is My Business*), mystery writer Raymond Chandler commented on this when he wrote "pulp fiction never dreamed of posterity." *Pulp* fiction was synonymous with *trash* fiction, and the nature of early pulp SF has been aptly described as "scientific pornography for the mechanically minded," and "writing which drooled over descriptions of technology."[2]

It was this sort of fiction that helped primed the imaginations of the millions who years later listened to Orson Welles' infamous 1938 dramatization (on his Radio Mercury Theatre program) of the 1898 novel *War of the Worlds* by H. G. Wells. In that Halloween eve, coast-to-coast broadcast, millions heard the horrifying news: Martians had invaded the planet, their first rockets landing in the little town of Grovers Mill, New Jersey! Hundreds were already said to be dead, and panic and terror swept the national listening audience. The near-hysterical public response to what had been merely a stunt so stunned the government that the FCC announced it would hold hearings on whether the "public trust" had somehow been violated. It was Gernsback's pulps, however, that had prepared massive numbers of people to seriously entertain the idea such an incredible event might even occur.

Gernsback's earlier publications, *Modern Electrics*, *Science & Invention* and *Radio News*, had printed SF from time-to-time, as had many of the 'ten-cent family magazines' since the 1890s. It was in *Modern Electrics* that Gernsback published, as a 1911 serial, his own historically important (but so incredibly awful that it's funny) tale "Ralph 124C 41+." Set in the year 2660, Ralph—a scientific genius who has a "gigantic mind"—pursues through space the villains who have kidnapped his sweetheart, Alice. There is no adversity in this tale that Ralph cannot overcome with the aid of some marvelous invention, created instantly on-the-spot using only (or so it seems) a bucket of whipped cream, an old garden hose, and a broken mousetrap. Even bringing Alice back from the apparently dead, in a perhaps unintentional imitation of the miracle described in the Gospel of John, where Jesus raises Lazarus from the dead, is not beyond Ralph's astonishing skill-set.

That sort of juvenile nonsense greatly declined in SF (although it didn't totally vanish) in later years, particularly after *Astounding Science Fiction* magazine and its editor John W. Campbell, Jr. (1910–1971) came on the scene in the late 1930s. With Campbell, who was editor[3] until his death, aspiring writers for

---

[2] Anthony Frewin, *One Hundred Years of Science Fiction Illustrations*, Jupiter Books 1974, p. 53.

[3] Campbell was also a writer, and his story "Who Goes There?" (which appeared in the August 1938 issue of *Astounding* under the pen-name of Don A. Stuart) is rightfully considered a masterpiece that straddles the horror and SF genres. The tale is of a shape-changing alien who terrorizes a scientific research team in the Antarctica; it has been filmed at least three times, most recently in 2012 as *The Thing*.

Astounding had to pay far more attention to the fundamental laws of nature than had been the case with Gernsback. *Astounding* still publishes today, under its new name (since 1960) of *Analog*. *Analog's* editors after Campbell have remained faithful to his commitment to science, and the magazine enjoys a reputation for publishing 'SF for engineers.' And it's still printed on pulp paper. Even before the pioneering pulps of Munsey and Gernsback, however, one can find the glimmerings of SF.[4]

2.2 Before the 'Start' of SF

Indeed, long before modern SF became populated with space aliens, intelligent robots, and time travelers, the extraordinary voyages of Jules Verne (*Journey to the Center of the Earth, From the Earth to the Moon, Around the World in 80 Days,* and *20,000 Leagues Beneath the Sea*) were the nineteenth century equivalent of science fiction (the difference between those two brilliant contemporaries, the Englishman Wells and the Frenchman Verne, is the difference between super-speculative science and super-high technology/engineering, respectively). More than a century before Verne and Wells, the 1735 *Gulliver's Travels* by Jonathan Swift (1667–1745) was an extraordinary voyage of the first-rank, and a century before *that* the German mathematician and astronomer Johannes Kepler (1571–1630)—best known today for the three laws of planetary motion named after him—had in 1611 written his posthumously published *Somnium* (*The Dream*) of a trip to the moon.

Perhaps less well-known is the equally imaginative 1638 work by the Anglican bishop Francis Godwin (1562–1633), *The Man in the Moone* (published posthumously), which describes a journey from the Earth to the Moon and back. Twenty years later (1657), Cyrano de Bergerac (1619–1655) did the same with his better known, posthumously published *L'Autre Monde: ou les États et Empires de la Lune (The Other World: or the States and Empires of the Moon).* In 1752 Voltaire turned the extraordinary voyage on its head, with Earth being visited by aliens from Sirius and Saturn in his *Micromégas.*

One of Swift's contemporaries broke completely free from the 'trip to the moon' theme[5] and its variations, and actually played with a modern SF idea—that

---

[4] Three useful guides to the pre-pulp SF literature are Marjorie Hope Nicolson, *Voyages to the Moon,* Macmillan 1948, Roger Lancelyn Green, *Into Other Worlds: space-flight in fiction, from Lucian to Lewis,* Arno Press 1975, and J. O. Bailey, *Pilgrims Through Space and Time: trends and patterns in scientific and utopian fiction,* Argus Books 1947 (reprinted in 1972 by Greenwood Press).

[5] A very old theme, in fact, as the second century AD Lucian of Samosata, in his *True History,* includes descriptions of a voyage to the moon and of interplanetary war. That was pretty far-out stuff 2,000 years ago.

of time travel. As observed by a present-day SF historian, "The first time-traveler in English literature is a guardian angel who returns with state documents from 1998 to the year 1728 in Samuel Madden's *Memoirs of the Twentieth Century*."[6] This premise was *slightly* improved upon a century-and-a-half later by Mark Twain in his *A Connecticut Yankee in King Arthur's Court* (1889), which used a knock on the head with a crowbar (instead of an angel) to achieve time travel.

Of course, just having a tale involving marvelous adventures isn't sufficient to make it an SF tale. Cervantes' *Don Quixote* (1605, 1615), Dumas' *The Three Musketeers* (1844), and the many old tales of Robin Hood and his Merry Men and of King Arthur and his Knights, are all simply bursting with adventures that are out of the ordinary, but nobody would call them SF. And of course all of these early efforts in marvelous events found ancient inspiration in exciting adventure story-telling in Homer's *Odyssey* and *Illiad*, and Virgil's *Aeneid*. I don't really think those are SF stories, either.

So, what *does* make a story an SF story? This question has prompted literary critics and analysts to write literally tons of papers and monographs, read mostly by other critics and analysts. The people who do that sort of thing tend to be professors of English, not SF writers (although there are, of course, some important exceptions, such as Gregory Benford and Stanislaw Lem). It has been my experience that many of the definitions such critics have come-up with can be problematical. One, for example, says that SF necessarily involves an unfolding future.[7] I think that *far* too restrictive, disqualifying many stories that I think clearly *are* SF.[8] Nevertheless, let's accept it and see where it might take us.

I'll start by quoting the fourth century BC Greek philosopher Aristotle, who wrote in his *Rhetoric* that "nobody can 'narrate' what has not yet happened. If there is narration at all, it will be of past events, the recollection of which is to help the hearers to make better plans for the future." This is an early statement of an ancient taboo against telling (or writing) a tale of the future, which certainly wouldn't have encouraged any potential SF authors in ancient Greece. Matters didn't change any time soon, either, as in one of his sermons the seventeenth century English poet and priest John Donne (1572–1631), who eventually became Dean of St. Paul's in London, declared "to write a chronicle of things before they are done" is "irregular" and "perverse."

---

[6] Paul Alkon, *Origins of Futuristic Fiction*, University of Georgia 1987, p. 85. Madden's work, more a satire than it is SF, was published in 1733. Madden was an Irish Anglican clergyman.

[7] Thomas A. Hanzo, "The Past of Science Fiction," in *Bridges to Science Fiction* (George E. Slusser et al., editors), Southern Illinois University Press 1980, p. 132.

[8] Just to name one; Isaac Asimov's beautiful 1958 short story "The Ugly Little Boy," the tale of a young Neanderthal boy plucked out of his time by present-day time machine experimenters. After studying their subject for a lengthy time, they grow weary of him and decide to send him back to the remote past, where he has no chance for a normal life and to the virtually certain fate of a quick, brutal death. Asimov's emotional ending will bring tears to all but the coldest of hearts.

The taboo against writing of the future actually makes some sort of theological sense in Donne's case, as doing that might well seem to a cleric to be mocking the religious prophecies of the Bible. *God's* words may speak of 'things to come,' but not those of mere men.

A real problem for the development of SF in ancient times (keeping the 'unfolding future' definition in mind) was that that the Bible itself didn't provide much time for either the future *or* the past. Early Christian theologians, who read the Bible as an historical document to be interpreted as the literal truth rather than as a literary device teaching lessons of moral behavior in the form of allegory, arrived at numerous, different dates for Creation. But while these dates were different, they did share the common feature of not being all that long ago. Martin Luther argued for 4000 BC, for example, and in agreement (but much more precise) was the Calvinist James Ussher, Archbishop of Armagh and Primate of All Ireland. Ussher declared "that from the evening ushering in the first day of the world, to that midnight which began that first day of the Christian era, there were 4003 years, seventy days, and six hours." He further asserted that Man was created on the sixth day, which was Friday, October 28. He didn't say anything, one way or the other, however, about the possible complications caused by leap years.

As eminent as these men were, there were others who thought they could do better, and so by the early nineteenth century there were more than 120 dates for Creation, spanning the interval 3616 BC to 6984 BC. Their one point of agreement was that the remote past really wasn't very remote. Similarly, Biblical prophecy of the coming final confrontation between good and evil—the Battle of Armageddon—and the Last Judgment didn't offer much of a lengthy future either. Theology just didn't give much 'time room' for SF adventurers. That all changed with the discovery of *geological time*, the discovery that Earth isn't a mere few thousand years old but rather is *billions* of years old. This realization, which began at the end of the eighteenth century, provided Charles Darwin with just what he needed for the theory of evolution in the 1859 publication of his *Origin of Species*; namely, a past of such vast duration—a *chasm* of time—so enormous as to stupefy biblical scholars.

You can't snicker at the scholars' reactions, as a billion years *is* just too much for most human minds to really grasp. It is truly humbling to historians to contemplate how very little of the past is known. As one anonymous wit once put it, "History is a damn dim candle over a damn dark abyss." A bit more scholarly was H. G. Wells, who in his 1944 doctoral thesis wrote "A thousand years is a huge succession of yesterdays beyond our clear apprehension." Some thinkers actually had the imagination to ask if the past might be *infinite* in extent, but others objected that if that were the case then everything would have already happened (!). Modern cosmologists think the Universe began about 15 billion years ago, with the famous Big Bang.[9] A finite past started by God does encourage some theologians to wonder what God was doing *before* the moment of Creation (probably, cynics reply, creating Hell for those who would ask such a question).[10]

In any case, if the Biblical accounts of *Genesis* could be called into account, then why not as well the Biblically limited future? Could the future be vast, too? Perhaps even unlimited? It is no coincidence that such a realization was in-step with the rise of a new literature of *time* adventures into the future; even before H. G. Wells' Time Traveller and his time machine, Edward Bellamy (1850–1898) had the protagonist in *Looking Backward* simply sleeping from 1887 into the future of the year 2000. Well, no matter how they got there, there was now a *place* for such temporal adventurers to go. As Alfred Tennyson wrote in his 1835 poem "Locksley Hall,"

> *"For I dipt into the future, far as human eye could see,*
> *Saw the Vision of the world, and all the wonder that would be"*

Today we label stories that speak of that Vision as science fiction.

2.3 Early Theological SF

It is always risky to state that some story is a 'first,' but I think there are two possibilities, the first not as strong as the second but still compelling enough to keep it in the running. It is the *Inferno*, the first part of the epic poem *The Divine Comedy* by Dante Alighieri (1265–1321). It is the story of Dante's descent into the earth (where most people who believe in Hell imagine its location) through the nine circles of Hell, with the Roman poet Virgil as his guide. Dante of course wrote more with theology and poetry than science in mind, and the journey is an allegory on the human soul's path to salvation and eventual eternal paradise with God—with *lots* of sinful temptations vividly described along the way.

---

[9] Just how brief is the length of mere human history is nicely illustrated by the so-called 'cosmic year.' If we imagine that the entire history of the Universe from the Big Bang to today is compressed into just 1 year, and that our present *now* is midnight of December 31, then dinosaurs were walking the Earth until the middle of yesterday, and Christ died on the Cross 4 s ago.

[10] This is not a joke, and quite serious modern thinkers continue to ponder the issue; see, for example, Brian Leftow, "Why Didn't God Create the World Sooner?" *Religious Studies*, June 1991, pp. 157–172.

The one 'scientific' aspect of *Inferno* occurs when Dante reaches the Earth's center, which is described as the *frozen* center of the Ninth Circle and *not* as the lake of fire and brimstone that has terrorized centuries of Sunday school kids. (It's amusing to note that a common saying for an event with no chance of occurring is that it will happen "when Hell freezes over." According to Dante, at least part of it already has!) Hell's center is where Satan is held in bondage as punishment for the ultimate sin of treachery against God, and where Dante discovers that gravity reverses direction. This is correct; I do find it curious, however, that Dante overlooked another interesting (and far more obvious) physical characteristic of the center, namely the immense pressure there. It is 'Hellishly high,' in fact, and you can only wonder at the additional, awful torments Dante could have delivered to sinners with it![11] Two present-day SF writers (Larry Niven and Jerry Pournelle) revisited *Inferno* in 1976, in a new take (with the same title) on Dante's journey.

For my second, and much stronger, candidate for the claim of being the first religious SF story, I offer the 1881 tale "Hands Off." It appeared under an anonymous by-line when *Harper's New Monthly Magazine* published it, but its author was the Unitarian minister Edward Hale (1822–1909), best known today as the author of the 1863 story "The Man Without a Country." Hale wrote "Hands Off" in an attempt to promote some theological debate (which it didn't), and it is certain he would be surprised to learn his tale is remembered today as a pioneer in SF. Hale was no neophyte in early SF as years earlier, in October 1869, *The Atlantic Monthly* had started publishing (as a serial) his "The Brick Moon," a tale of the first artificial satellite (a hollow 200 foot diameter sphere made of bricks).

"Hands Off" opens with the mysterious words "I was in another stage of existence. I was free from the limits of Time, and in new relations to space." These words are spoken by an unnamed narrator who seems to have just died and who finds himself, in his new 'form,' observing "some twenty or thirty thousand solar systems" while in the company of "a Mentor so loving and patient." Under the guidance of this Mentor (probably an angel), in attempt to 'improve' history, the narrator alters the Biblical account of Joseph and his imprisonment in Egypt on one of these systems.

At first, subsequent history *is* better, but then humanity sinks into irreversible depravity. In the end the narrator watches the last handful of humans kill each other at a particularly symbolic place for the Christian world: "The last of these human brutes all lay stark dead on the one side and on the other side of the grim rock of Calvary!" On this world there would be

---

[11] You can find an analytical treatment of Earth's interior gravity in my book *Mrs. Perkins's Electric Quilt*, Princeton 2009, pp. 186–214. In particular, on pp. 200–203 the pressure at the center of the Earth is calculated. (For the curious, it's 25,000 tons per square inch.)

no Crucifixion and Resurrection for the salvation of humankind, an outcome which naturally disturbs the narrator. But the Mentor calms him, saying "Do not be disturbed, you have done nothing." It has, you see, been just an experimental world, an alternate Universe, and so the narrator has learned the lesson of "Hands Off."

Hale's idea of a multitude of worlds created by God (of which ours is but one) sounds very much like the many-worlds view of reality that many find implicit in theoretical quantum mechanics. That view is a seemingly outrageous idea first put forth seriously *in science* by the physicist Hugh Everett III (1930–1982) in a 36-page, 1957 Princeton doctoral dissertation titled "On the Foundations of Quantum Mechanics." In the many-worlds interpretation, the entire Universe splits at the occurrence of every decision by every sentient being *everywhere* (on Earth, on the fourth planet orbiting the triple star system Rigel—if there is such an inhabited planet—, on *all* the inhabited planets in *all* of the galaxies, etc.), to always provide a distinct Universe for every possible sequence of decisions from The Beginning of Time to The End. Want to split the universe? Decide whether to blink your right eye or your left eye! (You can see why the word *outrageous* is used.) Outside of theoretical physics, the many-worlds concept had already appeared in an SF story—*without* Hale's theological nature—two decades before Everett, in Murray Leinster's 1934 tale "Sidewise in Time."[12]

The many-worlds idea had appeared *in art* 40 years before Hale, with almost certainly a theological twist, in a beautiful, fantastic illustration in the 1844 book *Un Autre Monde* (*Another World*). Known either as "The Infinity Juggler" or "The Juggler of Worlds," it was the work of the French artist Jean-Ignace Isidore Gérard (1803–1847), who published under the name 'Grandville.' The juggler—Grandville's version of Hale's Mentor—appears as a court jester who is clearly having fun manipulating his multitude of worlds, while the man (humanity?) in the foreground watches. The man appears to be simultaneously fearful and fascinated, involved yet clearly impotent. Is Earth one of the worlds among which the Jester stands, or is it one of those flying through space? Or is Earth, perhaps, simply the unfortunate world ingloriously stuffed down the front of the Jester's pants? (That surely would explain a lot!) If born a hundred years later, Grandville would have easily found work as an artist in the imaginative world of the SF pulps.

---

[12] 'Murray Leinster' was the pen-name of William F. Jenkins (1896–1975).

Grandville's Infinity Juggler

The idea of a multitude of worlds, expressed in Grandville's art and in Hale's story, continued to fascinate long after their appearance. In a short essay (it really isn't a real *story*) by James Gunn (born 1923) called "Kindergarten," for example, we learn that God made the Solar System as a school assignment when just a youngster in his kindergarten (!) class. He is the *slowest* youngster in the class, in fact. The piece ends with an on-going argument between the Teacher who is clearly an entity beyond God (whatever that might mean) and His parents (whatever that might mean) on whether or not to destroy Earth as a flawed effort. The theology in all this is more than just a little bit beyond curious, to be sure, but I include Gunn's effort here because it is a multi-world creation tale published in a relatively recent pulp (the April 1970 issue of *Galaxy Magazine*). But I can't help but wonder—just what are the *other* kids in that kindergarten class now doing?

Years later, as another example, in the much deeper 2000 novel *Calculating God* by Robert Sawyer (born 1960), we learn that God has been an experimenter in evolution when multiple alien civilizations discover that all the great historical extinctions experienced on Earth also occurred *at the same time* on their worlds, as well. So, here we have God as scientist who apparently learns-as-He-goes from experience—pretty much like the rest of us!

Returning to the start of the twentieth century, the idea of time travel (which I'll discuss in far more detail in Chap. 7) was already afoot in early SF, mostly because of Wells' *Time Machine*. Although a pioneering work in time travel fiction, *The Time Machine* contains essentially no discussion about the consequences of paradoxes, the heart-and-soul of this sub-genre of SF. The closest Wells comes is during the dinner party, in the opening of the story, when the Time Traveller attempts to convince his friends of the possibility of a time machine. One of them observes that such a gadget "would be remarkably convenient for the historian. One might travel back and verify the accepted account of the Battle of Hastings, for instance." To that another guest replies "Don't you think you would attract attention? Our ancestors had no great tolerance for anachronisms." The Time Traveller has no reply to that (because, I think, *Wells* had no reply).

It didn't take long for another writer to fill that gap, however, with the 1904 publication of *The Panchronicon* by Harold MacKaye (1866–1928). An Edwardian time machine with style, the Panchronicon is a large container that swings, on a rope tether, around a steel pole erected at the North Pole. By "cutting the meridians" faster than does the sun, it (and its occupants) travels through space and time from 1898 New Hampshire to the London of three centuries earlier.[13]

In the course of Mackaye's novel we follow the adventures of the time travelers as they encounter such puzzles as changing the past and meeting yourself, situations that would receive a great deal of attention from SF writers in years to come. The often made, incorrect assertion common in early SF (even into the 1930s), that a backward-moving time traveler would grow ever younger, is refuted. Most impressive of all, I think, is the novel's clever treatment of an *information loop in time*. Specifically, we learn how a Shakespeare who is bedeviled by writer's block nevertheless came to write one of his plays: one of the time travelers simply whispers the magic words she has memorized (for her literary club meetings) into his ear. Does this make Shakespeare a plagiarist (of himself)? More to the point, however, is *this* question: in whose brain were those 'magic words' created? This is a question that still excites a lot of debate among physicists and philosophers.

---

[13] MacKaye might have been inspired to use this idea from a reading of Edgar Allen Poe's 1841 story "Three Sundays in a Week," in which a bit of amusing turmoil is caused by a character moving across time zones. That story is, however, not SF by any interpretation. (Poe's 1835 "The Unparalled Adventure of One Hans Pfaall," of a voyage to the moon, is a better candidate and I'll say more about it in Chap. 6 when we discuss aliens in SF.)

Before MacKaye, and even Wells, other writers tried to find additional twists to time travel. One alternative to simply invoking a *time machine* with which to observe the past was to imagine a faster-than-light rocket (this was before the theory of special relativity said you can't do that); with such a rocket one might, at least in principle, look backward in time by traveling out into space and then watch the light from the past that your high-speed trip had outrun. The French astronomer Camille Flammarion (1842–1925), for example, had made this dramatic idea a centerpiece in his 1887 novel *Lumen*, which describes how a man just dead (in 1864) instantly finds his spirit on the star Capella where he watches the light then arriving from the Earth of 1793 bearing images of the French Revolution.[14]

By the beginning of the twentieth century the idea of watching the past by outrunning light had drifted down into juvenile literature. For example, the French writer Jean Delaire (1888–1950) used this idea of outrunning light in her 1904 novel *Around a Distant Star*, in which a man builds a spaceship that can travel at 2,000 times the speed of light. With it he and a friend travel to an Earth-like planet 1,900 light-years distant and then use a super-telescope to watch the Crucifixion and resurrection of Jesus.

The use of religion in SF made a dramatic appearance in the July 1939 issue of *Astounding Science Fiction*, the pulp edited by John W. Campbell, Jr. The story "Trends" was the first sale by Isaac Asimov to that magazine, and it concerned the imagined social *resistance* the builders of the first moon rocket might experience. Asimov later recalled that it wasn't the moon trip itself that fascinated Campbell (that was an old, much used idea by 1939), but rather the idea of religious opposition to space travel. In the story the leader of the Twentieth Century Evangelical Society (as well as the League of the Righteous) declares, any such attempt to leave Earth would be "profaning the heavens" and to "defy God." The only reward the rocketeers would receive would be "Divine vengeance." That is because "It is not given to man to go wheresoever ambition and desire lead him. There are things forever denied him, and aspiring to the stars is one of these. Like Eve [the rocketeers wish] to eat of the forbidden fruit, and like Eve [they] will suffer due punishment therefor."

The story reads as "incredibly naïve" (Asimov own words) today, now that we know just how difficult it is to build a moon rocket. The first rocket is sabotaged, but an eventually successful second rocket is built in secret by a handful of men in the backwoods of northern Minnesota. And all ends well when the religious leader dies, the rocket trip succeeds, and the rocketeers are acclaimed to be national heroes. Although technically simplistic to the point of being a juvenile

---

[14] As a point of fact, Capella is 42 light-years from Earth, which is at odds with the 71 years between the man's death and the French Revolution. You'd think an *astronomer* wouldn't make a mistake like this!

fantasy, "Trends" was nevertheless a daring story, too, one that risked condemnation from powerful religious organizations that could easily have taken offense at being portrayed as irrational to the point of committing violence. That didn't happen—perhaps because religious leaders didn't read SF!—but still, it was a gamble that both Asimov and Campbell took.

Two years after "Trends" appeared, Asimov was inspired by Campbell to write another story with a very strong religious nature to it. It would be a near-parody of Biblical prophecy; it would be, in fact, what Asimov felt was the best piece of short fiction that he ever produced. Asimov recalled years later that the premise for "Nightfall" originated in Campbell's mind when, during a visit by Asimov to *Astounding*'s offices in March 1941, Campbell read a quotation from Ralph Waldo Emerson: "If the stars should appear one night in a thousand years, how would men believe and adore; and preserve for many generations the remembrance of the city of God ...!"[15] Those words prompted Campbell to ask Asimov what he thought would happen if the stars actually did appear only for brief times after long intervals of absence. When Asimov had no reply, Campbell gave him the story idea: men would go mad.

The way Asimov set this interesting idea into story form was to imagine a technical civilization on the planet Lagash, which is in orbit around a cluster of six stars. Nobody on Lagash has ever seen the night sky, as there is always at least one sun above the horizon. SF writers have found such multiple-star systems intriguing, perhaps in part because the orbit of a planet in the complicated, ever evolving gravitational field of a star-cluster would be highly convoluted, offering lots of interesting story angles. In Stanislaw Lem's novel *Solaris*, for example, the planet Solaris is in a double-star system; the only 'inhabitant' of the planet is its mysterious ocean, which seems to have the ability to stabilize what would otherwise be a highly variable orbit. (I think Asimov's *six*-star system holds the SF record!)

In "Nightfall" Lagash's *eight*-body (remember, the *total* system consists of the planet, the six stars, and the Lagash's moon) orbit[16] is such that, every 2,049 "years" (what a "year" is on the planet, in Earth-years, is not given), the only star of the six that is in the sky at that time is eclipsed by Lagash's moon. That moon has not actually been observed because the "eternal blaze of the two [major] suns ... drown it out completely." Its existence is suspected, though, because the

---

[15] The quotation is from the opening of Emerson's essay *Nature*, written in 1836. "The city of God" is a clear reference to the universe, itself.

[16] To appreciate just how wild this planet's orbit could be, see the discussion of mere *three*-body orbits in my book *Number-Crunching*, Princeton 2011, pp. 131–217. A nice exposition on the calculation of the 'habitable zone' (water exists in liquid form) in a simple two-star system is by Su-Shu Huang, "Life-Supporting Regions in the Vicinity of Binary Systems," *Interstellar Communication* (A. G. W. Cameron, editor), W. A. Benjamin 1963, pp. 93–101.

observed orbit of Lagash is not in accordance with the inverse square law of gravitation. (The physical scientists of Lagash are apparently at the same stage of development as were Earth scientists at the start of the twentieth century; that is, in possession of Newton's theory of gravity but not of Einstein's general theory of warped spacetime).

Further calculations using the inverse square law have shown that theory and observations can be brought into agreement with the additional presence of a supposed moon, and *that if it exists this moon will soon produce an eclipse.* Asimov's clever idea was to have these calculations motivated by a *Book of Revelations*, central to a religious cult on Lagash. That *Book* contains the story of something mysterious called the "Stars" and, as one character explains, "The Cultists said that every two thousand and fifty years Lagash entered a huge cave, so that all the suns disappeared, and there came *total darkness all over the world!* And then, they say, things called Stars appeared, which robbed men of their souls and left them unreasoning brutes, so that they destroyed the civilization they themselves had built up. Of course they mix all this up with a lot of religio-mystic notions, but that's the central idea."

The Fifth Chapter of the *Book of Revelations* describes what happens in some detail once the 'cave' is entered (Asimov, who had a reputation for being a pretty irreverent fellow, must have had a lot of fun writing *this*!):

"And it came to pass that in those days [one sun] held lone vigil in the sky for ever longer periods as the revolutions passed; until such time as for full half a revolution, it alone, shrunken and cold, shone down upon Lagash. And men did assemble in the public squares and in the highways, there to debate and to marvel at the sight, for a strange depression had seized them. Their minds were troubled and their speech confused, for the souls of men awaited the coming of the Stars. And in the city of Trigon, at high noon, Vendret 2[17] came forth and said unto the men of Trigon, 'Lo, ye sinners! Though ye scorn the ways of righteousness, yet will the time of reckoning come. Even now the Cave approaches to swallow Lagash; yea, and all it contains.' And even as he spoke the lip of the Cave of Darkness passed the edge of [the sun] so that to all Lagash it was hidden from sight. Loud were the cries of men as it vanished, and great the fear of soul that fell upon them. It came to pass that the Darkness of the Cave fell upon Lagash, and there was no light on all the surface of Lagash. Men were even as blinded, nor could one man see his neighbor, though he felt his breath upon his face. And in this blackness there appeared the Stars, in countless numbers, and to the strains of music of such beauty that the very leaves of the trees cried out in wonder. And in that moment the souls of men departed from them, and their abandoned bodies became even as beasts; yea, even as brutes of the wild; so that through the blackened streets of the cities of Lagash they prowled with wild cries. From the Stars there then reached down the Heavenly Flame, and where it touched, the cities of Lagash flamed to utter destruction, so that of man and of the works of man nought remained."

This all sounds quite mysterious, of course, until the inverse square law calculations explain it in terms of mathematical physics. That doesn't mean all is okay, however, because as one story character says "This is not the century to preach 'The end of the world is at hand' . . . You have to understand that people don't believe the *Book of Revelations* anymore, and it annoys them to have scientists turn about face and tell us the Cultists are right after all—." To that a scientist replies "While a great deal of our data has been supplied us by the Cult, our results contain none of the Cult's mysticism. Facts are facts, and the Cult's so-called mythology *has* certain facts behind it. We've exposed them and ripped away their mystery." The religious sect isn't at all happy about this development of a scientific explanation for their *Book*. As the sect's leader complains to the scientists, "Your pretended explanation backed our beliefs, and at the same time removed all necessity for them. You made of the Darkness and of the Stars a natural phenomenon and removed all its real significance. That was blasphemy." In other words, having a 'mystery' is preferred over having an explanation, a condition that many would argue is not at all uncommon today.

What is particularly unnerving about the Fifth Chapter is that it nicely fits together with the current archaeological theory that says Lagash's history has a cyclic nature. As one scientist explains it, "This cyclic character is—or rather, was—one of the great mysteries. We've located series of civilizations, nine of them definitely, and indications of others as well, all of which have reached heights comparable to our own, and all of which, without exception, were destroyed by fire at the very height of their culture. And no one could tell why. All centers of culture were thoroughly gutted by fire, with nothing left behind to give a hint as to the cause."

There would always be a few who would survive each such calamity, of course, and that would further explain the origins of the *Book of Revelations*, itself. As we are told by one of the scientists, "The very insensitive would be scarcely affected—oh, such people as some of our older, work-broken peasants. Well, the children would have fugitive memories, and that, combined with the confused, incoherent babblings of the half mad morons, formed the basis for the *Book of Revelations*. Naturally, the book was based, in the first place, on the testimony of those least qualified to serve as historians; that is, children and morons; and was probably edited and re-edited through the cycles." (As with "Trends," Asimov took a real chance at offending powerful, established religious institutions with this transparent mocking of their Holy Books but, as before, he got away with it.)

And so Lagash is plunged into total darkness, the stars come out for as long as the eclipse lasts and, as Campbell wanted, Asimov has everybody on the planet go insane. Perhaps with good reason, too, as we learn that "Lagash was in the center of a giant cluster. Thirty thousand mighty suns shone down in a soul-searing splendor that was more frighteningly cold in its awful indifference than the bitter wind that shivered across the cold, horribly bleak world." As Asimov's last paragraph eerily describes the start of the eclipse (and the start of the next cycle), "The awful splendor of the indifferent Stars leaped nearer to them. On the horizon outside the window, . . . a crimson glow began growing, strengthening in brightness, that was not the glow of a sun. The long night had come again." (This really strikes me as a glaring—no pun intended—weak-point in the story because if Lagash is in the center of such a massive star cluster the night sky would actually be pretty bright and the surface of Lagash would not be at all dark.)

Well, forget *my* reservations; the fact is that editor Campbell loved "Night-fall" and so did his magazine's readers (the story appeared in the September 1941 issue of *Astounding*). Part of the fun readers had was being *in* on an inside-joke Asimov had woven into his tale. Near the end of it one scientist says he has developed a "really cute notion" about what the *Book*'s reference to "Stars" might be all about. As he explains, "Well, then, supposing there were other suns in the universe. I mean suns that are so far away that they're too dim to see. It sounds as if I've been reading some of that fantastic fiction, I suppose. . . . During an eclipse, these dozen suns would become visible because there'd be no *real* sunlight to drown them out. Since they're so far off, they'd appear small, like so many little marbles. Of course the Cultists talk of millions of Stars, but that's probably exaggeration. There just isn't any place in the universe you could put a million suns—unless they touch one another. . . . And I've got another cute little notion. Have you ever thought what a simple problem gravitation would be if only you had a sufficiently simple system? Supposing you had a universe in which there was a planet with only one sun. The planet would travel in a perfect ellipse and the exact nature of the gravitational force would be so evident it could be accepted as an axiom. Astronomers on such a world would start off with gravity probably before they even invented the telescope. Naked eye observation would be enough."

When asked if a 'one planet, one sun' system would be stable, he replies "Sure! They call it the 'one-and-one' case. It's been worked out mathematically, but it's the philosophical implications that interest me. . . . Of course, there's the catch that life would be impossible on such a planet. It wouldn't get enough heat and light, and if it rotated there would be total Darkness half of each day. You couldn't expect life—which is fundamentally dependent upon light—to develop under those conditions." One of his friends tries to be supportive, saying that even though all that *is* pretty crazy stuff, still "It's nice to think about as a pretty abstraction—like a perfect gas, or absolute zero." Asimov's intent with including this little exchange in the story was, of course, so *Astounding*'s readers could condescendingly smile to themselves with their 'superior knowledge' that such a thing *is* possible.

Looking back at the early days of magazine SF, one modern author and critic could write "I used to moan over the fact that pulp magazines were printed on pulp paper and steadily decompose back towards the primordial from which they sprang. I am beginning to feel that this is a bit of a good thing."[18] Asimov (who had a doctorate in chemistry from Columbia University), however, had based "Nightfall" on real, solid science, and that story (as well as those of Robert Heinlein that Campbell was also starting to publish in *Astounding*) showed that the critic who once described early magazine fiction as "science that was claptrap and fiction that was graceless"[19] had to admit, as SF moved into the 1940s and more modern times, that things were definitely starting to look-up.

Of course, in 1941 there were still a few rough spots in "Nightfall." For example, early in the story one character says, to show that the sophisticated elite weren't being taken-in by either the Cultists or the scientists, "Investors don't really believe the world is coming to an end, but just the same they're being cagy with their money until it's all over. Johnny Public doesn't believe you, either, but the new spring furniture might just as well wait a few months—just to make sure." It's simply astonishing how '1940s, New York City wise-guy-like' that inhabitant on far-away Lagash sounds—but you can't *really* have expected SF pulp to have completely changed overnight. Don't forget, when he wrote "Nightfall" Asimov was just 21 years old, as well as that he lived in New York City and already had a reputation for being just a bit of a wise-guy himself.

There were other hurdles, too, for early SF to jump, with a really big one being what to do about girls and sex. Sexual behavior, in particular, attracts the attention of religious theoreticians and, since young men are a major fraction of the SF readership, this is not a trivial issue. Of early pulp SF, Anne McCaffrey (1926–2011), a highly successful SF author, wrote the following in a hilariously funny essay: "Prior to the '60s, stories with any sort of love interest were very rare. True, it was implied in many stories of the '30s and '40s that the guy married the girl whom he had rescued/encountered/discovered

---

[18] Harry Harrison, "With a Piece of Twisted Wire . . .," *SF Horizons* 1965 (no. 2), pp. 55–60.
[19] See the editors' introduction to *Famous Science-Fiction Stories* (R. J. Healy and J. F. McComas, editors), Random House 1957.

during the course of his adventures. But no real pulse-pounding, tender, gut–reacting scenes. The girl was still a 'thing' to be 'used' to perpetuate the hero's magnificent chromosomes. Or perhaps, to prove that the guy wasn't 'queer.' I mean, all those men locked away on a spaceship for months/years at a time. I mean . . . and you know what I mean even if I couldn't mention it in the sf of the '30s and '40s."[20]

Later, when we get to Chap. 6 and the possibility (or not) of interstellar space travel to meet alien beings, one of the stories discussed is Robert Heinlein's 1941 "Common Sense." It is set on a so-called 'generational spaceship' in which generation after generation of people are born, live their lives, and die as the ship makes its enormously long voyage to a distant star. (The story makes the implicit admission that faster-than-light travel a' la *Star Trek* is not possible.) In that society men (even if clearly morons) are the 'natural' superiors of women, and the physical abuse of women (including getting teeth knocked out) when they need 'discipline' is described as being acceptable. Providing even more support for McCaffrey's thesis of how shabbily some early SF treated women is the description in the story of women being 'natural' physical cowards while men (even if clearly morons) are uniformly brave. One critic, commenting on the 1930s pulps that specialized in romance stories for young women, observed that the heroes and heroines in such tales often displayed the "mental equipment of a banana split."[21] That would not have been a valid characterization for the majority of the science-oriented readers of pulp SF, but the fact that Heinlein published the sometime cartoonish "Common Sense" in *Astounding Science Fiction* magazine shows that he clearly appreciated the occasionally socially immature teenage male audience for which he was writing. Heinlein was prone, too, to stroking the often inflated egos of his young readers by implying that *they*, as readers of science fiction, also understood actual science better than did the readers of those 'other' pulps.

In "Common Sense," for example, he wrote "space ship ballistics is a very simple subject, being hardly more than the application of the second law of motion to an inverse-square field. That statement runs contrary to our usual credos; it happens to be true." No, that *isn't* true, as anyone who has actually worked through the mathematical physics of the 'mere' three-body problem (of calculating the orbits of three massive bodies, with each moving in the combined fields of the other two) soon comes to appreciate.[22] Perhaps, however, my criticism is a bit unfair to Heinlein as he was, after all, in the business of

---

[20] See "Hitch Your Dragon to a Star: Romance and Glamour in Science Fiction," in *Science Fiction, Today and Tomorrow* (Reginald Bretnor, editor), Harper & Row 1974, pp. 278–292.

[21] Margaret MacMullen, "Pulps and Confessions," *Harper's Monthly Magazine*, June 1937, pp. 94–102.

[22] See my book *Number-Crunching*, Princeton 2011, pp. 131–217.

telling a good story, not that of teaching science. And perhaps some of his readers, intrigued by his casual dismissal of difficult topics, were intrigued enough to study them and, in fact, to become real scientists.

2.4 Theological Maturity

Two literary events, as the decade of the 1950s came to its end, showed the world beyond science fiction that SF could provide deep, serious treatments of religion. These were the appearances of the 1958 novel *A Case of Conscious* by James Blish, and the novel *A Canticle for Leibowitz* by Walter M. Miller, Jr. (1923–1996) which appeared the very next year. Both novels were the result of combining several linked short stories that their authors had published a few years earlier in the pulps (Miller in the *Magazine of Fantasy & Science Fiction*, and Blish in *IF: Worlds of Science Fiction*). Both are today recognized as classics (each won the prestigious SF Hugo award for best novel of the year), with Miller's using only the first of the common elements of SF (space travel, aliens, or some fantastic gadget like a time machine), while Blish's makes use of the first two. I'll discuss Miller's book (which, unlike much of religious SF, is quite sympathetic to the Church) here, and *A Case of Conscious* later in the book (Chap. 6).

The 1959 *A Canticle for Leibowitz* opens in a post-apocalyptic world in the American desert, six centuries after a nuclear war (called the "Fire Deluge") has destroyed much of civilization. It is a science fiction sequel to Nevil Shute's 1957 *On the Beach* (assuming people had survived that novel's world-wide atomic radiation). When Brother Francis, a novice at the remote Leibowitz Abbey accidently discovers the ruins of an ancient fallout shelter that contains a human skull with a gold tooth (a skull that glows in the dark from residual radiation), he begins to suspect he is into something extraordinary. He soon *knows* that is the case when, in the ruins of the shelter, he stumbles upon a rusted box containing numerous strange items, including an ancient paper bearing the words "CIRCUIT DESIGN BY: *Leibowitz, I. E.*" He has discovered legendary relics of the beatified founder of his religious Order, Isaac Edward Leibowitz!

After the Fire Deluge, murderous mobs of simpletons who had survived the war begin to kill scientists, teachers, and technicians, along with all the other educated people they can find, people that the mindless hordes have decided deserve to die for having helped to destroy the world. It is a reign of terror called "The Simplification." For the literate, the only available escape is the Church, which takes them in, vests them in monk's robes, and hides them away in monasteries and convents. Many are thus protected, but many others are still discovered; the fate of those poor unfortunates is either to burn in a fire or to hang at the end of a rope. Isaac Edward Leibowitz manages to avoid both of those grim outcomes for some years, and he becomes a priest and founds his new Order with the blessing of the Church. But, in the end, he is betrayed[23] and he, too, dies a martyr's death at the end of a strangulation noose while hanging over a fire.

Along with people, the Church has attempted to preserve human history and knowledge, much as it did in the Dark Ages. The material it gathers becomes known as 'The Memorabilia,' and while it soon fades into being beyond understanding to the monks, the preservation of it all is a sacred mission. The monks will honor that duty, if required, for the next 10,000 years. Before then, it is hoped, a means for rediscovering the secrets of The Memorabilia will appear. Besides securing what original books they can find, the monks of Leibowitz Abbey hand-copy them, too, illuminating algebra texts with "cheerful cherubim surrounding tables of logarithms," and faithfully reproducing blueprints of electrical apparatus right down to every detail (including what might only be "the stain of a decayed apple core" left accidently on the diagram by some long-dead draftsman).

The rest of the novel, as the centuries pass, follows both the Order of Leibowitz and the Church. More than a thousand years after Brother Francis' find of the fallout shelter, humanity has again gone full circle. Leibowitz, long since canonized, is the patron saint of electricians, and the technology of nuclear weapons has been rediscovered. However, a repeat of the Fire Deluge has been avoided long enough for the secret of the interstellar starship drive to be discovered and so, when atomic war does again threaten, escape is possible.

The novel ends as the horror of nuclear war erupts once again, but this time the Church is ready: it has anticipated the coming disaster and has activated contingency plans for escape. As "the horizon became a red glow" and "the visage of Lucifer mushroomed into hideousness," monks and sisters board themselves and children into a starship. As the last monk to enter the ship pauses at the hatchway before sealing it shut, he looks at the glow in the sky and says *Sic transit mundus* ("thus passes the world"). And then the starship thrusts itself heavenward, like the collective soul of all humanity departing a corpse, towards salvation somewhere beyond Earth.

---

[23] The reader slowly learns, as the novel progresses, that Leibowitz had been a weapons scientist (perhaps an electrical engineer) who had sent his wife (who had a gold tooth) to the shelter under the pretext of carrying secret documents to safety. In reality, those documents where nothing but routine papers; it was all just a ruse to convince her to seek shelter.

Chapter 3
Time, Space, God's Omniscience, and Free Will

3.1 What Is Time?

As you might expect, there are those who look at a question like the one above as an opportunity for some jest. One such wit answered it with "Time is just one damn thing after another," and another (perhaps the same person) thought "Time is what keeps everything from happening at once." Amusing, sure, but we'll try for something just a little bit deeper than that in this chapter!

Christian clerics had identified time as something unusual long before SF writers and their time travel stories. We can, in fact, trace the theological interest in time back in time (no pun intended) at least fifteen centuries, to St. Augustine who, in his *Confessions* wrote "What, then, is time? I know well enough what it is, provided that nobody asks me; but if I am asked what it is and try to explain, I am baffled." Certainly the seventeenth century Spanish Jesuit Juan Eusebius Nieremberg caught the spirit of wonder that time holds for the devout when he wrote, in his *Temperance and Patience* that "Time is a sacred thing; it flows from Heaven ... It is an emanation from that place, where eternity springs ... It is a *clue* cast down from Heaven to guide us... It hath some assimilation to Divinity."

Going outside of Christianity, we can find equally strong reactions to the mystery of time. From Plutarch's *Platonic Questions* we learn that when the question of time's nature was put to Pythagoras he simply uttered the mystical "time is the soul of this world." The *Laws of Manu* of Hinduism, the *Torah* of Judaism, the *Koran* of Islam, and the revealed truths of Gautama Buddha are all full of references to time. It is, in fact, to the pagan gods of Greek mythology that we owe our 'modern' image of Chronos, or Father Time.

Not just the Greeks made time a god. In the *Bhagavad Gita* (*Song of the Lord*), the central religious-romantic epic of Hinduism that predates Christ by five centuries, one of the characters reveals his divine nature and declares his power

P.J. Nahin, *Holy Sci-Fi!*, Science and Fiction,
DOI 10.1007/978-1-4939-0618-5_3, © Springer Science+Business Media New York 2014

thus: "Know that I am Time, that makes the worlds to perish, when ripe, and bring on them destruction." And in the even more ancient Egyptian *Book of the Dead*, which dates back over three thousand years, the newly deceased was thought to literally become one with time itself. The merging of time and the resurrection of the body after death is demonstrated in the line "I am Yesterday, Today and Tomorrow, and I have the power to be born a second time."

Lovely words, yes, but they don't really tells us what time *is*. Einstein felt that, according to his general theory of relativity, time and space would cease to exist if the universe was empty, which is in step with one of his favorite philosophers, Spinoza. In his *Principles of Cartesian Philosophy*, Spinoza declared "there was no Time or Duration before Creation."

3.2 Time in SF and Theology

The mystery of time was directly addressed by R. H. Hutton (1826–1897), a Unitarian minister and the literary editor of the English journal *The Spectator*, when he wrote[1] in an 1895 review of Wells' *Time Machine* that "the story is based on that rather favorite speculation of modern metaphysicians which supposes *time* to be at once the most important of the conditions of organic evolution, and the most misleading of subjective illusions . . . and yet Time is so purely subjective a mode of thought, that a man of searching intellect is supposed to be able to devise the means of traveling in time as well as in space, and visiting, so as to be contemporary with, any age of the world, past or future, so as to become as it were a true 'pilgrim of eternity.'"

Novelist Israel Zangwill (1864–1926) wrote a much more analytic review of his friend's novel for the *Pall Mall Magazine*; he was the only Victorian reviewer to attempt a scientific analysis of the concept of traveling through time. Although he thought Wells' effort was a "brilliant little romance," he also thought the very concept of a time machine to be "much like the magic carpet of *The Arabian Night*," and was far more enthusiastic about flying faster than light—in that way, he wrote, one could watch "the Whole Past of the earth still playing itself out." As mentioned in the last chapter, Flammarion had already used this idea in fiction, and Delaire would again a few years after Zangwill's review.

Decades later Stanislaw Lem would declare these two ideas, time travel and FTL (faster than light) travel, to be "very convenient inventions" that were part of "a bastard of myths gone to the dogs" which had "domesticated the cosmos for story telling purposes" to the point that SF "has lost its strange, icy sovereignty."[2]

---

[1] Reprinted in Patrick Parrinder, *H. G. Wells: The Critical Heritage*, Routledge & Kegan 1972.
[2] Stanislaw Lem, "Cosmology and Science Fiction," *Science Fiction Studies*, July 1977, pp. 107–110.

Even though Lem used both of those 'inventions' a lot in his own writing, I assume from his declaration that he felt there was a big price—a fall from reality into fantasy—to be paid for invoking them. And he didn't back down from that position when challenged by fellow SF writer and physicist Gregory Benford.[3] Although Benford didn't specifically mention either invention, Lem made it a point in his rebuttal[4] to write "My point was the 'holistic' falsification of . . . the real universe by SF; to wit, the irreversibility of the time arrow [that is, time travel] and the impossibility of faster-than-light travel." He claimed the SF use of time travel and FTL "short-circuit" the real universe. (Despite this strong criticism, however, Lem often used both devices in his own stories!)

A different way from FTL travel to look backward in time is found in the idea that time is a 'closed loop' and so, to see the past, all we need do is simply look *forward* sufficiently far! That is, time curves back on itself. This was the view of Plato (circa 400 BC), for example, and it was actually a reasonable interpretation of what he observed in nature, with the seemingly endless repetition of the seasons, the regular ebb and surge of the tides (the old English word *tid* is, in fact, a unit of time), the unvarying alternation of night and day, and the travels of the planets around their closed orbits in the sky. Whatever might be observed today, it seemed obvious to Plato, would happen again in the future. This view of time has an ancient suggestive visual symbol, the Worm Ouroborous, or World Snake that eats its own tail endlessly.

Plato's most famous student, Aristotle, held the same circular view of time. Aristotle believed the world had already traveled around 'time's loop' forever and so was infinitely old. The closed circularity of time was a central image in his mind; as he dramatically put it, in circular time it is equally true that we live both after *and* before the Trojan War. This is not just a fantasy idea from the remote past. Closed, circular time was favorably mentioned, for example, by the famed modern-day theoretical physicist Stephen Hawking in his well-known book *A Brief History of Time*. Hawking's position there is that with circular time there is no need for God since there is no first event—and so no need for a 'First Cause!'[5] (A little aside: Modern cosmological thought is that the universe started with the famous Big Bang, whose 'cause' was . . . well, we can only speculate. There *are* of course those who *do* think they have the answer (God), but of course there are others who remain unconvinced. They ask the obvious next

---

[3] Gregory Benford, "On Lem on Cosmology and SF," *Science Fiction Studies*, November 1977, pp. 316–317.
[4] Stanislaw Lem, "In Response to Professor Benford," *Science Fiction Studies*, March 1978, pp. 92–93.
[5] You can find some interesting commentary on Hawking's 'theology' in the book by the English philosopher Antony Flew (1923–2010), *There Is a God: How the World's Most Notorious Atheist Changed His Mind*, HarperCollins 2007. Flew is the former atheist in the title who dramatically announced in 2004 that had 'found God.'

question of 'What caused God?', with the answer usually of the form 'nothing caused God, as God needed no cause.' That leaves unaddressed the puzzle of why one couldn't simply say that of the Big Bang, itself.)

Returning to Hawking, there even seems to be Biblical support for circular time in *Ecclesiastes* 1:9: "The thing that hath been, it is that which shall be; and that which is done is that which shall be done; and there is no new thing under the Sun." The modern, popular view of time (no matter what Hawking may actually believe, or what the Bible may say), however, is *linear* time. That is, straight-line time extending backwards into the past and forward into the future (and, to no doubt misapply Kipling just a bit, "never the twain shall meet"). This is the view you'll most commonly find in modern time travel SF.[6] It is somewhat ironic to note that it was the *Christian theological doctrine* of unique historical events that gave rise to the linear time that even the most atheistic modern physicist surely accepts as 'obvious.' The Creation of the world and of Adam and Eve, the adventures of Noah and his Ark in the famous cataclysmic Flood, the Death and Resurrection of Jesus—these were all to be interpreted as events that occurred in sequence, *once*. None would happen again, and so for Christianity circular time just would not do.

In addition, a central spiritual aspect of Christianity (and a significant reason for its huge appeal to the common folk of Jesus' time even as they encountered brutally harsh Roman suppression) was that it introduced the *expectation of change* into the static world of ancient times. With Christianity, people and all their children's children didn't have to be, *always and forever*, impoverished toilers doomed to miserable, short lives in which they had no power to alter their fate. The future could be *different*—perhaps worse, yes, but also perhaps better.

3.3 The Four-Dimensional World

In various SF stories the world is imagined as being four-dimensional. A world constructed, that is, from the three spatial dimensions we all directly experience *plus* either one additional and mysterious *space* dimension that is somehow hidden from direct observation, *or* those three spatial dimensions plus

---

[6] This is not always the case; for example, Asimov's 1956 *non*-time travel story "The Last Question", to be discussed later in this book, is based on circular time. And I have to mention, too, the strange 1967 novel *Counter-Clock World* by Philip K. Dick (1928–1982) in which time runs backward (buried people come alive again and emerge from their graves as the "Sacrament of Miraculous Rebirth" is intoned by priests). Reversed time is an old fantasy, in fact, as you can find it in Plato's 360 BC dialogue *Statesman*. It has continued to fascinate modern authors as well, including those we normally don't think of as writing SF; see, for example, the famous 1922 story "The Curious Case of Benjamin Button" by F. Scott Fitzgerald (1896–1940).

time. Of course, in the first interpretation we would still have time as well and so a four *spatial* dimensional world would in fact be *five-* dimensional![7] In physics, mathematics, *and* in SF, spaces of any dimension (whether their nature be space *or* time) beyond the first three spatial ones are called *hyperspaces*. Both the space and the time views have had theological uses in SF.

The idea of space as the fourth dimension can be traced back to Aristotle who, writing in 350 BC, declared in his essay "On the Heavens" that "the three dimensions are all that there are." Centuries later, the second-century AD. Greek astronomer, Ptolemy, argued the same. But even the pronouncements of these profound thinkers didn't end such speculations. In 1878, for example, the well-known Scottish mathematical physicist Peter Tait (1831–1901) wrote that a fourth spatial dimension might offer one way to explain such otherwise inexplicable occurrences as ghosts and the reading of sealed letters. (Perhaps this was the inspiration for Oscar Wilde's 1887 short story "The Canterville Ghost," where we read that the ghost in the title, at one point, makes a retreat and disappears through the wainscoting by "hastily adopting the Fourth Dimension of Space as a means of escape.")

In an 1875 book he co-authored (*The Unseen Universe*), Tait speculated that a human soul might be a four-dimensional knot (!) in the ether (ether is 'stuff' once thought to fill the universe and through which light waves could 'wave,' and which modern physics has long since abandoned as imaginary). The association of the fourth dimension with the spirit world can be traced as far back as the mid-seventeenth century and the Cambridge philosopher Henry More (1614–1687). Two centuries later the idea of two parallel worlds—ours and another inhabited by the spirits of the dead that share a common time dimension but each with three spatial dimensions displaced along a fourth spatial dimension) was used by Elizabeth Phelps (1844–1911) in her best-selling 1868 novel *The Gates Ajar*. Her novel *might* have been included in the previous chapter as an example of 'early theological SF' but I decided to defer mention of it until now. That's because Phelps didn't write with the intention of telling a 'great adventure' tale, but rather to offer ease from the terrible emotional pain suffered by the millions who had lost loved ones in the violence of the American Civil War—and who had found little if any comfort for their loss in traditional religions.

H. G. Wells also used this same idea in his 1895 (the same year *The Time Machine* appeared in book form) novel *The Wonderful Visit*. That non-SF

---

[7] One of Superman's more interesting adversaries in the comics of the 1940s and 1950s was Mr. Mxyzptlk (pronounced *mix-yez-pittle-ick*), a being with seemingly magical powers who was from the Land of Zrfff in the fifth dimension. His powers weren't really magic, of course, but resulted solely from his extra-dimensionality. The comics aren't traditional SF, but some come pretty close and they have always been pulp.

work describes the adventures of an angel who flies into 'our' world where he is shot in the wing by a Vicar's gun. All is quickly 'explained' with passing mention of the fourth dimension: "There may be any number of three dimensional Universes packed side-by-side," that are "lying somewhere close together, unsuspecting, as near as page to page in a book." He used the spatial interpretation of the fourth dimension in others of his SF writings, as well; see, for example, Wells' 1897 novella *The Invisible Man*, and the short stories "Davidson's Eyes" and "The Plattner Story."

Before the end of the nineteenth century at least two 'non-fictional' religious books appeared that interpreted hyperspace as the dwelling place of God Himself: Alfred Taylor Schofield's 1888 *Another World*, which declared God's hyperspace to be of four spatial dimensions, and Arthur Willink's 1893 *The World of the Unseen* which took the even bolder leap into a divine hyperspace with an *infinity* of spatial dimensions (what a mathematician would today call a *Hilbert space*, after the great German mathematician David Hilbert (1862–1943)).

The idea of time, rather than space, as the fourth dimension is much more current these days. It is just as old an interpretation, however, as it can be traced back to the eighteenth century.[8] Still, it wasn't until a curious letter appeared in the British scientific journal *Nature* in 1885 that the view of time as the fourth dimension was mentioned in a serious way.[9] The author, mysteriously signing himself only as "S.," began by writing "What is the fourth dimension? ... I [propose] to consider Time as a fourth dimension ... Since this fourth dimension cannot be introduced into space, as commonly understood, we require a new kind of space for its existence, which we may call time-space." Who was S.? Nobody knows, but Professor Bork (note 8) speculates that it was an acquaintance of H. G. Wells.

It is most likely that S. was *not* Wells, himself, and in support of that we have a near-denial from him. In his 1934 *Experiment in Autobiography* he wrote "In the universe in which my brain was living in 1879 there was no nonsense about time being space or anything of that sort. There were three dimensions, up and down, fore and aft and right and left, and I never heard of a fourth dimension until 1884 or thereabout. Then I thought it was a witticism."

Others besides Professor Bork have speculated on S.'s identity[10] but, as far as SF is concerned, time became the popular view as the fourth dimension

---

[8] A. M. Bork, "The Fourth Dimension in Nineteenth Century Physics," *Isis*, October 1964, pp. 326–338.

[9] S., "Four-Dimensional Space," *Nature*, March 26, 1885, p. 481.

[10] See, for example, Bernard Bergonzi, *The Early H. G. Wells: a study of the Scientific Romances*, University of Toronto Press 1961, pp. 31–32. When I wrote to the editorial offices of *Nature* about S., I was informed that the journal's archives contain no clue as to S.'s identity.

with the publication of Wells' masterpiece *The Time Machine*. The novella opens with "The Time Traveller" expounding on a recondite matter to a group of his friends. As he asserts, "There is no difference between Time and any of the three dimensions of Space except that our consciousness moves along it." When asked to say more about the fourth dimension, he replies, "It is simply this. That Space, as our mathematicians have it, is spoken of as having three dimensions, which one may call Length, Breadth, and Thickness, and it is always definable by reference to three planes, each at right angles to the others. But some philosophical people have been asking why *three* dimensions particularly—why not another direction at right angles to the other three?—and have even tried to construct a Four-Dimensional geometry. Professor Simon Newcomb was expounding this to the New York Mathematical Society only a month or so ago."[11]

One has to be careful not to jump from those words in an SF classic to the conclusion that Wells possessed some sort of hidden insight into the possible physics of a time machine. Wells, in fact, clearly stated that, for him, *The Time Machine* was 'just' a really good story. Indeed, as he wrote in the 1934 preface to a new edition of the novella, time as the fourth dimension was for him simply a "magic trick for a glimpse of the future."

3.4 The Block Universe

One of the repercussions of viewing time as the fourth dimension, one with profound implications for both SF and theology, was expressed in a manner strangely reminiscent of S.'s letter to *Nature*. In 1920 another cryptic note, signed this time as "W. G." (as with S., *Nature* has no record of who W. G. was), appeared, containing the following provocative passage:

> "Some thirty or more years ago a little *jeu d' esprit* was written by Dr. Edwin Abbott entitled *Flatland* . . . Dr. Abbott pictures intelligent beings whose whole experience is confined to a plane, or other spaces of two dimensions, who have no faculties by which they can become conscious of anything outside that space and no means of moving off the surface on which they live. He then asks the reader, who has consciousness of the third dimension, to imagine a sphere descending upon the plane of Flatland and passing through it. How will the inhabitants regard this phenomenon? They will not see the approaching sphere and will have

---

[11] And so he actually was. Simon Newcomb (1835–1909) was an eminent American astronomer and, in 1897–1898, President of the American Mathematical Society. Wells read Newcomb's December 28, 1893 Address to the New York Mathematical Society when it was reprinted in *Nature*, February 1, 1894, pp. 325–329.

no conception of its solidity. They will only be conscious of the circle in which it cuts their plane. This circle, at first a point, will gradually increase in diameter, driving the inhabitants of Flatland outward from its circumference, and this will go on until half the sphere has passed through the plane, when the circle will gradually contract to a point and then vanish, leaving the Flatlanders in undisturbed possession of their country ... Their experience will be that of a circular obstacle gradually expanding or growing, and then contracting and they will attribute to *growth in time* what the external observer in three dimensions assigns to a movement in the third dimension. Transfer this analogy to a movement of the fourth dimension through three-dimensional space. Assume the past and future of the Universe to be all depicted in four-dimensional space, and visible to any being who has consciousness of the fourth dimension. If there is motion of our three-dimensional space relative to the fourth dimension, all the changes we experience and assign to the flow of time will be due simply to this movement, *the whole of the future as well as the past existing in the fourth dimension* [my emphasis]."[12]

W. G.'s words are a clear statement of what is called the *block universe* view of four-dimensional spacetime, a view of reality as a once-and-forever entity. This idea long-predates W. G., however, as we can find it stated by Wells in *The Time Machine* 25 years earlier. From the previous section, you'll recall that the novella opens with the Time Traveller introducing a group of friends to the idea of time as the fourth dimension, and in that same speech he says "There is no difference between Time and any of the three dimensions of Space except that our consciousness moves along it ... here is a portrait of a man at 8 years old, another at 15, another at 17, another at 23, and so on. All these are evidently sections, as it were, Three-Dimensional representations of his Four-Dimensional being, *which is a fixed and unalterable thing* [my emphasis]."

But we can go back even further in time, *much* further than Wells' 1895 story, all the way back to the fifth-century BC and the words of the Greek philosopher Parmenides. His view of reality: "It is uncreated and indestructible; for it is complete, immoveable, and without end. Nor was it ever, nor will it be; for now it *is*, all at once, a continuous *one*." As an echo of this nearly early 2,000 years later, in the thirteenth century *Compendium Theologiae* of Thomas Aquinas, we read "We may fancy that God knows the flight of time in His eternity, in the way that a person standing on top of a watchtower embraces in a single glance a whole caravan of passing travelers."

And in his *Summa Theologiae* Aquinas wrote "Now although contingent events come into actual existence successively, God does not, as we do, know

---

[12] W. G., "Euclid, Newton, and Einstein," *Nature*, February 12, 1920, pp. 627–630.

them in their actual existence successively, but all at once; because his knowledge is measured by eternity, as is also his existence; and eternity which exists as a simultaneous whole, takes in the whole of time … Hence all that takes place in time is eternally present to God." Somewhat paradoxically, however, Aquinas *did* make a distinction between past and future because, in that same work, he declares that "God can cause an angel not to exist in the future, even if he cannot cause it not to exist while it exists, or not to have existed when it already has." For Aquinas, then, whereas the past is rigid and unchangeable, the future is plastic, and these are *not* characteristic features of the block universe view of reality.

The block universe is a *fatalistic* universe, the one described in Omar Khayyam's eleventh century poem *The Rubaiyat* with these words: "And the first Morning of Creation wrote, What the Last Dawn of Reckoning shall read." Those very words were quoted to the students of the Harvard Divinity School in a March 1884 address made by the Harvard psychologist William James, in a talk titled "The Dilemma of Determinism." That title is a bit misleading, however, as James argued for free-will, which *is* allowed in a deterministic world (but not in a fatalistic one). Determinism says 'If you do A then B will happen, but if you do not do A then (perhaps) something other than B will happen.' Free will is not excluded in a deterministic world because you are *free to choose* to either do A or not to do A.

A fatalistic world (like the block universe), on the other hand, simply says either 'You *will* do A' or 'You *will not* do A' and which path is your path is *not* your choice. Two years before his Divinity School talk James had *really* unloaded on the fatalistic block universe, calling it a world that had "the oxygen of possibility all suffocated out of its lungs," and one in which "there can be neither good nor bad, but [only] one dead level of mere fate."[13]

No matter James' emotional rejection, the *analytical* Einstein fully embraced the block universe view of reality. In a letter dated March 21, 1955 that he wrote to the children of one of his dearest friends who had recently died, he said "And now he has preceded me briefly in bidding farewell to this strange world. This signifies nothing. *For us believing physicists, the distinction between past, present, and future is only an illusion, even if a stubborn one* [my emphasis]"[14] This last sentence must have been more than just a bit enigmatic to his friend's children, and years later a nice elaboration of it was given in the technical physics literature:

---

[13] William James, "On Some Hegelisms," *Mind*, April 1882, pp. 186–208. The 'Hegel' in the title is the German philosopher Georg Wilhelm Friedrich Hegel (1770–1831)—whose endorsement of the block universe James greatly disliked—saying somewhat harshly of the German: "Hegel's philosophy mingles mountain-loads of corruption with its scanty merits."

[14] Banesh Hoffmann, *Albert Einstein: Creator & Rebel*, New American Library 1972, pp. 257–258.

"It seems that Einstein's view of the life of an individual was as follows: If the difference between past, present, and future is an illusion, i.e., the four- dimensional spacetime is a 'block Universe' without motion or change, then each individual is a collection of myriad of selves, distributed along his history, each occurrence *persisting on the world line*,[15] *experiencing indefinitely the particular event of that moment* [my emphasis]. Each of these momentary persons . . . would possess memory of the previous ones, and would therefore believe himself identical with them; yet they would all exist separately, as single pictures in a film."[16]

As I've already mentioned, the block universe view had been around in philosophy and theology for a very long time before Einstein, and it had actually appeared in fiction even before Einstein's birth. The block universe is implied, for example, near the end of the 1842/3 story "The Mystery of Marie Rogêt" by Edgar Allen Poe, where we read "It is not that the Deity *cannot* modify his laws, but that we insult him in imagining a possible necessity for modification. In their origin these laws were fashioned to embrace *all* contingencies which *could* lie in the Future. With God all is *Now*."

Three decades later, in "The True Story of Bernard Poland's Prophecy" by George Eggleston (1839–1911), which is about a man who sees his own coming death in the yet-to-occur American Civil War (the tale appeared in the June 1875 issue of *American Homes* magazine), there is the following passage where Bernard speaks to an unnamed friend, the narrator:

"Do you know," said Bernard, presently, "I sometimes think prophecy isn't so strange a thing . . . I really see no reason why any earnest man may not be able to foresee the future, now and then . . ."

"There is reason enough to my mind," I replied, "in the fact that future events do not exist, as yet, and we can not know that which is not, though we may shrewdly guess it sometimes . . ."

"Your argument is good, but your premises are bad, I think," replied my friend, . . . his great, sad eyes looking solemnly into mine.

"How so?" I asked.

"Why, I doubt the truth of your assumption, that future events do not exist as yet . . . Past and future are only divisions of time, and do not belong at all to

---

[15] A *world line* is the trajectory of a point in the four-dimensional spacetime of the block Universe. This imagery is due not to Einstein, but rather to Herman Minkowski (1864–1909), who was Einstein's mathematics professor during his student days in Zurich. Minkowski first described the world line concept in 1908, in his famous *geometrical* interpretation of Einstein's 1905 *mathematical* theory of special relativity.

[16] L. P. Horwitz, *et al.*, "On the Two Aspects of Time: the distinction and its implications," *Foundations of Physics*, December 1988, pp. 1159–1193.

eternity . . . To us it must be past or future with reference to other occurrences. But is there, in reality, any such thing as a past or a future? If there is an eternity, it is and always has been and always must be. But time is a mere delusion . . . *To a being thus in eternity, all things are, and must be present. All things that have been, or shall be, are* [my emphasis]."

Eggleston was not intentionally writing SF, but with that passage he came pretty close to it (as well as to the same view that Einstein would later adopt in his 1955 letter).

Bernard Poland's words "being thus in eternity" are commonly thought to refer to God, and it requires a further extension of 'ordinary' four-dimensional spacetime. As one character in the short-story "The Time Conqueror" by Lloyd Eshbach (1910–2003), which appeared in the July 1932 issue of the pulp magazine *Wonder Stories*, says, "Beyond the fourth there is a fifth dimension . . . Eternity, I think you would call it. It is the line, the direction perpendicular to time." This might be where Aquinas would have imagined 'God's watchtower' (that I mentioned earlier in this section) to be located. Not going quite so far as to put *God* there, in Isaac Asimov's 1955 novel *The End of Eternity* the 'time police'[17] oversee the endless centuries from a place outside of time called Eternity.

The block universe, itself, specifically appeared in quite early pulp SF magazine fiction; one example of that is the short story "The Machine Man of Ardathia" by George Weiss (1898–1946)—writing under the pen-name "Francis Flagg"—in the November 1927 issue of *Amazing Stories*. There a time traveler from the future, and a man of the present (the narrator), have the following exchange:

"I have just been five years into your future."
"My future!" I exclaimed. "How can that be when I have not lived it yet?"
"But of course you have lived it."
I stared, bewildered.
"Could I visit my past if you had not lived your future?"

God's eternity, and the nature of His relationship to spacetime, is *not* clear from the Bible. For example, consider the Old Testament story of King Ahab (*First Kings* 21). Ahab coveted Naboth's vineyard but Naboth wouldn't sell. The King retreated, but his wife Jezebel arranged for Naboth's downfall and judicial murder and thus caused the arrival of all his property into her husband's hands.

---

[17] In SF the *time police* are charged with preventing time travelers from changing the past, either purposefully or by accident. Attempts to change the past are a popular device in theological SF, and I'll say more on that topic later in the book.

This angered God, who commanded Elijah to prophesy disaster on Ahab's house. A then fearful Ahab responded with sackcloth *and at that God shifted the predicted disaster to the house of Ahab's son* ... the point here is that God, declared to be omniscient, seems to have been surprised at Ahab's penitence!

God is indeed aware of everything in this Biblical tale, *but only as it happens*; that is, God's knowledge is subject to growth. And so we see that the ancient Hebrew concept of God as a participant in history is at extreme odds with the present-day Christian conception that God's divine knowledge is of all that has been, all that is, and all that will be. Just like the Old Testament view of omniscience, the present-day view of divine eternality also has Biblical support. (Finding support in the Bible for opposing claims is nothing new, of course.) For example, "For I am the Lord, I change not" (*Malachi* 3:61) and "the Father ... with whom is no variableness" (*James* 1:17).

3.5 God's Omniscience in Theology and SF

The omniscience of God is fundamental to all the major theistic religions, including Christianity, Judaism, and Islam. Such divine foreknowledge, however, would appear to be in direct conflict with free-will in humans, a belief that the same religions, in an apparent contradiction, fully embrace. That conflict was described by Geoffrey Chaucer's *Troilus and Criseyde*, a poem written in the Middle Ages more than 600 years ago:

> Some say "If God sees everything before
> It happens—and deceived He cannot be—
> Then everything must happen, though you swore
> The contrary, for He has seen it, He."
> And so I say, if from eternity
> God has foreknowledge of our thought and deed,
> We've no free choice, whatever books we read.

The claim made in Chaucer's poetry denying free-will was repeated by Wells in *The Time Machine*, or at least it was in the original print appearance of the story. Before it appeared in book form, the novella was serialized in the *New Review*, and in that magazine debut there is a passage in the Time Traveller's speech to his friends connecting omniscience and the block universe, a passage that Wells, for some reason, deleted from the later book:

> "I'm sorry to drag in predestination and free-will, but I'm afraid those ideas will have to help ... Suppose you knew fully the position and properties of every particle of matter, of everything existing in the Universe at any particular

moment of time: suppose, that is, that you were omniscient. Well, that knowledge would involve the knowledge of the condition of things at the previous moment, and at the moment before that, and so on. If you knew and perceived the present perfectly, you would perceive therein the whole of the past. If you understood all natural laws the present would be a complete and vivid record of the past. Similarly, if you grasped the whole of the present, knew all its tendencies and laws, you would see clearly the future. To an omniscient observer there would be no forgotten past—no piece of time as it were that had dropped out of existence—no blank future of things yet to be revealed . . . [P]resent and past and future would be without meaning to such an observer . . . He would see, as it were, a Rigid Universe filling space and time . . ."

Wells' "Rigid Universe" certainly sounds like the block universe, and he seems to have believed that it held important implications for the concept of free-will. Nevertheless, while the question of free-will does arise regularly in many time travel stories, the fact is that it does not appear in the final form of *The Time Machine*.[18]

Wells almost surely got the idea for this doomed passage from reading (or reading about) the famous "powerful intellect" imagined in 1814 by the French mathematical physicist Pierre-Simon Laplace (1749–1827). That year, in the Introduction to his *Essai philosophique sur les probabilités*, he wrote that if such an intellect knew the position and velocity of every particle at a given time, along with the laws of nature, then it could calculate the position of any particle at any other time. Imagine, he asked,

"An intellect which at a certain moment would know all forces that set nature in motion, and all positions of all items of which nature is composed; if this intellect were also vast enough to submit these data to analysis, it would embrace in a single formula the movements of the greatest bodies of the universe and those of the tiniest atom; for such an intellect nothing would be uncertain and the future just like the past would be present before its eyes."

Laplace's claim was abandoned with the development of quantum mechanics at the start of the twentieth century, in particular the Heisenberg Uncertainty Principle, which says such wonderful knowledge is intrinsically impossible to obtain.

So, physics seems to deny us a natural way to achieve omniscience, but of course God is *super*natural and so theological SF *can* legitimately make use of

---

[18] For this passage in the *New Review* version of the story, see *The Definitive Time Machine: A Critical Edition of H. G. Wells' Scientific Romance with Introduction and Notes* (H. M. Ceduld, editor), Indiana University Press 1987, pp. 176–177.

this interesting ability. Arthur C. Clarke did that in at least two of his short stories, "The Nine Billion Names of God" (1953) and "The Star" (1955). In both tales the ability is not specifically commented on, but for each to make any sense requires that God be omniscient.

The first story opens with a Lama from a Tibetan monastery purchasing a Mark V Automatic Sequence Computer from a New York firm. When asked just what a remote monastery high in the lonely mountains of Tibet wants with the latest computer, the answer is a simple if surprising one: it is to speed-up a project the Lama's lamasery has been working on for 300 years—the printing of all possible names of God. Without the computer and doing it all by hand, the job will take another 15,000 years, and that is just too long to wait. With the aid of the computer, however, the job can be finished in just 100 days.

The analytical nature of the task is easy to understand. As the Lama explains, "All the many names of the Supreme Being—God, Jehovah, Allah, and so on—they are only man-made labels." Believing that the real name of the Supreme Being is no longer than nine characters in an unspecified alphabet, the computer will be programmed to systematically print all character string permutations of that alphabet with no string longer than nine characters. (As an added complication, no character can repeat in succession more than three times.) Somewhere in that list will be all the real names of God. No matter how strange a task this may seem, with the transfer of funds from the Lama's substantial account at the Asiatic Bank the deal is struck.

The story then jumps forward in time to Tibet where we listen-in to a discussion between George and Chuck, two computer engineers the Lama hired to operate the Mark V. Chuck, it seems, has just learned the *real* reason behind the search for all of God's real names. As he explains to George, the priests "believe that when they have listed all His names—and they reckon that there are about nine billion of them[19]—God's purpose will be achieved. The human race will have finished what it was created to do, and there won't be any point in carrying on. Indeed, the very idea is something like blasphemy." When George asks 'what happens then?' the answer is: "When the list's completed, God steps in and simply winds things up . . . bingo!" In other words, it's the end of the world.

---

[19] That is, there are nine billion real names of God, embedded in the vastly larger number of all possible character strings. To get a feel for just how large is that number, suppose the special alphabet used by the priests has 26 characters (just like English). Then, if we don't worry about the restriction that there be no run of a given character longer than 3, there are a total of $26 + 26^2 + 26^3 + \ldots + 26^9$ strings, a geometric series easily summed to give $\frac{26^{10} - 26}{25} \approx 5.6 \times 10^{12}$. That is, five *trillion* strings plus six hundred billion *more*. When the run restriction is applied, the number of possible strings is of course reduced from this, but we are still left with a *lot* of strings. I think Clarke grossly overestimated the ability of any 1950s computer (as well as underestimating the amount of paper required to print all those strings)!

With both men now understandably concerned, they arrange matters so that, just before the computer finishes its computations, they are already out of the monastery and on their way to the airplane that is waiting to fly them back to civilization. They don't want to still be anywhere in the vicinity of the priests when nothing (so they believe) happens as the final string permutation is printed. As the two approach the plane, Chuck looks upward and, shocked at what he sees, tells George to look, too. The final sentence of the story is chilling: "Overhead, without any fuss, the stars were going out."

The required omniscience on God's part is, of course, due to the finite speed of light. Many (if not nearly all) of those stars must have been extinguished by God long ago, long before King Tut as born (and certainly long before the priests even began their work three centuries before), in order for their synchronized vanishing to appear on Earth *just as* the Mark V finishes its job and when George and Chuck look towards the heavens. Apparently Clarke liked this idea, as he used it again just 2 years later in his "The Star." You'll recall from the opening section of Chap. 1 that H. G. Wells wrote a short story in 1899 with this same title, in which humankind learned that the laws of nature care not a twit, one way or the other, about humankind. In his story, Clarke seems to be asking the opposite question: 'What if God cared too much?'

"The Star" opens in approximately the year 2540,[20] with a Jesuit science officer struggling with a deep personal crisis while on-board an interstellar spaceship. He is the chief astrophysicist on a scientific mission returning from the Phoenix Nebula, which is located "at the very frontiers of the explored universe," 3,000 light-years from the Vatican. The mission was to study the star that, when it exploded 6,000 years ago into the supernova that created the Nebula, had reduced itself to a white dwarf, an exotic object "smaller than Earth, yet weighing a million times as much."

We are told, as the story begins, that something happened during the mission that has profoundly shaken the priest, something so shocking as to make him question his faith. So severe is this shock that it has opened a fundamental crack in what has been the bedrock of his spiritual existence. The very motto of The Society of Jesus (AD MAJOREM DEI GLORIAM, "for the greater glory of God") mocks him. The story then slowly unfolds to reveal to the reader what has so disturbed the scientist-priest.

As the spaceship had approached the white dwarf at the start of the mission, an automatic search was started for any surviving planets. Any planet at an

---

[20] We know this as we are told it has been a thousand years since the founding of The Society of Jesus (which was in the year 1540). As an aside, even though 600 years have passed the ship's computer is just a Mark VI, only one generation beyond the computer in "The Nine Billion Names of God." (This *is* a silly quibble, of course, as the world *ended* in Clarke's earlier story!)

'ordinary' distance from the original star would have been vaporized and blown away like a hand-full of flour in a puff of wind, but to the expedition's surprise one planet *was* detected. It was at such an immense distance from the dwarf that it would have been the star's Pluto. Even at its great separation from the star, the surface of the small world had been horribly blasted but the world, itself, had not been obliterated.

And on that world, still intact, they had discovered the Vault.

An advanced civilization had flourished on an inner planet and, knowing from their scientists of their approaching extinction, the people of that lost world had constructed the Vault on the most remote planet in their system. Into the Vault they had placed thousands of visual records of their time on the cosmic stage, in the hope that those records would eventually be found and so at least a memory of what they had accomplished would survive.

One of those records, in particular, haunts the Jesuit: "One scene is still before my eyes—a group of children on a beach of strange blue sand, playing in the waves as children play on Earth. ... And sinking into the sea, still warm and friendly and life-giving, is the sun that will soon turn traitor and obliterate all this innocent happiness." But haunting the priest even more than that image is the question he has been asked by others in the expedition who are not believers: "It is one thing for a race to fail and die, as nations and cultures have done on Earth. But to be destroyed so completely in the full flower of its achievement, leaving no survivors—how could that be reconciled with the mercy of God?"

And at least as tortuous for the Jesuit is his discovery that this very supernova had already been seen centuries before, *on Earth*. As the expedition team speeds home at faster-than-light speed, he thinks "I know how brilliantly the supernova whose corpse now dwindles behind our speeding ship once shone in terrestrial skies. ... Yet, oh God, there were so many stars you could have used. What was the need to give these people to the fire, that the symbol of their passing might shine above Bethlehem?"

As Clarke points out in his story, in the Milky Way galaxy alone a hundred stars explode each year as a nova, and three or four times every one thousand years or so our galaxy produces a supernova. So, perhaps, the Bethlehem star was just a coincidence, and had no religious significance. But if it wasn't a random occurrence, then why would a god deserving of worship have used *that particular* star for what was, after all, not much more than what a new car dealership would do to announce itself with searchlights shining in the sky? If God actually did sacrifice a world to a stellar oven, then how could He escape being called a mass murderer?[21] In any case, if we assume it was the intention

---

[21] This harsh characterization wouldn't have shocked someone like Samuel Clemens (Mark Twain), who greatly disliked all organized religion: as he famously declared, "If there is a God, he is a malign thug."

of God to use that supernova as a 'searchlight' to guide the three Wise Men (the Magi in the Gospel of Matthew) to the birth of Jesus, then God must have initiated that remote star's detonation thousands of years *before* the birth of Jesus, and so His omniscience is again implied by the finite speed of light.

Neither of Clarke's two tales really makes omniscience a central point, leaving that 'talent' to be only faintly implied in each by a final, punch line. In contrast, the human implications of omniscience are front-and-center in "The Weed of Time," a 1970 story by Norman Spinrad (born 1940). The narrator, born in 2040 and who dies in 2150, is a man who is omniscient because he chewed the leaves of a plant brought back to Earth by the first expedition to the Tau Ceti system. Called *tempis ceti* (or just Temp), something in the leaves has a strange effect on all who eat them. As the Captain of the expedition describes the animals on the fifth planet of Tau Ceti, "they all seem to be herbivores and they seem to live off one species of plant which dominates the planetary flora. No predators. And it's not hard to see why [as] all the critters seem to know what the other animals will do before they do it." (By some sort of astonishing prescience—ironically appropriate in a story on omniscience!—Tau Ceti, a real star a 'mere' 12 light years from Earth, has recently been discovered to *actually have* a planetary system of five planets, with the outermost one indeed lying in the star's habitable zone.)

Omniscience might seem to be a wonderful ability, at first glance, but Spinrad's narrator gives us a quite different take on it: "For me, time as you think of it does not exist. I do not move from moment to moment sequentially like a blind man groping his way down a tunnel. I am at all points in the tunnel simultaneously ... *I am trapped in this eternal hell and I can never escape* [my emphasis], not even into death. My life is immutable, invariant, for I have eaten of Temp, the Weed of Time!" At one point in the story the narrator compares his life with the chapter of a book: the chapter is of finite length, with a fixed beginning and a fixed end, and yet within the book the entire chapter exists 'eternally.'

When he is interrogated by an operative from a government agency about future events, he does tell the man what he wants to know—but all the while he thinks to himself "I know that it is no use trying to tell ... them that knowledge of the future is useless, that the future cannot be changed because it was not changed because it will not be changed. They will not accept the fact that *choice* [my emphasis] is an illusion caused by [moving] along the timestream one moment after the other in blissful ignorance. They refuse to understand that moments of future time are no different from moments of past or present time: fixed, immutable, invariant. They live in the illusion of sequential time."

Spinrad's use of this of a rigid future is a statement of what philosophers sometimes call the *Master Argument* (a name referring to its supposed invulnerability to rebuttal), which can be traced back to the *Discourses* of the first century AD Roman philosopher Epictetus:

(a) The future follows from the past;
(b) And the past is unchangeable;
(c) And clearly what results from the unchangeable is, itself, unchangeable;

It therefore follows that the future is unchangeable.

Thousands of years later both the English poet William Blake (1757–1827) and the English scientist Oliver Lodge (1850–1940) were still wondering about this very issue. Blake was the more subtle, hinting at the puzzle of time in his epic poem "Jerusalem: The Emanation of the Giant Albion" with the words

"I see the Past, Present & Future, existing all at once
Before me, O Divine Spirit sustain me on thy wings!"

Lodge was far more direct when he bluntly asked "Is the future all settled beforehand, and only waiting to be 'pushed through' into our three-dimensional ken? Is there no element of contingency? No free will? I am talking geometry [of four-dimensional spacetime], not theology."[22]

In Spinrad's tale *omniscience* means the block universe, with free choice simply an illusion. The block universe is a 'book of destiny,' and the story's 'message' seems to be that omniscience is a power best left to the ken of God alone as it brings only the 'eternal hell of simultaneous awareness' to a mere human. Despite Lodge's words, theology *does* have a very big stake in the truth (or not) of the fatalistic block universe. The universe of Spinrad's tale, with its fixed future, gives rise to the question of why should we bother agonizing over the many decisions each of us makes every day? If the future is fixed, then we shouldn't bother and Christian theologians are left with the puzzle of explaining the meaning of the Biblical exhortation in *Deuteronomy* 30:19: "I call Heaven and Earth to record this day against you, that I have set before you life and death, blessing and cursing; therefore *choose* [my emphasis] life, that both thou and thy seed may live."

Theologians had, long before the rise of SF, been uncomfortably aware of the potential for conflict that a belief in free-will brings with it. As I mentioned earlier, Christianity, Islam, and Judaism all suppose God to be omniscient *and also omnipotent*. Let's now follow where that takes us. When He made humans he either did or did not give them free-will. If He did, then it follows that He cannot control the acts of humans—which forces us to conclude that He is *not* omnipotent. On the other hand, if He did not give humans free-will then the

---

[22] Oliver Lodge, "The New World of Space and Time," *Living Age*, January 24, 1920, pp. 240–244.

only way God can escape being responsible for the evil humans do is to suppose that He didn't give humans free-will because he couldn't—which means He is *not* omnipotent. So, no matter what God did concerning free-will He ends-up being *not* omnipotent, a conclusion in conflict with an all-powerful God.

I opened this chapter with a quotation from St. Augustin's *Confessions*, about his puzzlement over the nature of time. To complete this chapter, let me quote some more of his words, ones that followed those I gave you at the start:

> "I confess to you, Lord, that I still do not know what time is. Yet I confess too that I do know that I am saying this in time, that I have been talking about time for a long time, and that this long time would not be a long time if it were not for the fact that time has been passing all the while. How can I know this, when I do not know what time is? Is it that I do know what time is, but do not know how to put what I know into words? I am in a sorry state, for I do not even know what I do not know!"

St. Augustine's lament over the puzzle of time is the science fiction writer's challenge and, in a later chapter on time travel and theology, we'll see just how well SF has responded to that challenge.

Chapter 4
Religious Robots

4.1 Alan Turing, Artificial Minds, and the Souls of Machines

To speak of a religious robot might seem to be maximum silliness at the least, and downright blasphemous at the most, and so let me start this chapter with a little history before we get into the theological SF. In 1936 the English mathematician Alan Turing (1912–1954) started what is today called 'computer science.' He didn't actually call it that—the very first electronic digital computer was still almost a decade in the future—but Turing was a genius and he nevertheless quickly realized that a possible goal for the theoretical framework he had created was the eventual construction of an artificial (non-human) intelligence (AI). That is, the creation of an intelligent *robotic brain*.[1] Turing's AI work was parallel in time with the famous contributions of the MIT mathematician Norbert Wiener (1894–1964): Wiener's 1948 book *Cybernetics*, and then the 1950 book *The Human Use of Human Beings: Cybernetics and Society* in which he warned of the possible misuses of automata. Years later, in 1964, came his short work *God and Golem, Inc.*, in which he commented "on certain points where cybernetics impinges on religion." But it was Turing, not Wiener, who directly and enthusiastically embraced the concept of a *thinking* machine.

Turing believed that AI could eventually result in machines that would be indistinguishable from humans in terms of cognitive reasoning; and then later even surpass humans on that score. Turing's importance to the on-going work

---

[1] Turing's pioneering 1936 paper "On Computable Numbers, . . ." is not easy reading. But, if you want to give it a try, a good place to start is with the following book-length expansion of what Turing wrote, with *each line* by Turing expanded into a quite long (and quite good) explanation: Charles Petzold, *The Annotated Turing: a guided tour through Alan Turing's historic paper on computability and the Turing Machine*, Wiley 2008.

P.J. Nahin, *Holy Sci-Fi!*, Science and Fiction,
DOI 10.1007/978-1-4939-0618-5_4, © Springer Science+Business Media New York 2014

today in AI research by computer scientists is illustrated by the attachment of his name to a procedure, described in 1950, by which it might one day be determined if that goal has been achieved: the so-called *Turing test* (although he called it the *imitation game*).[2]

Turing's motivation in developing his game was the problem posed by the question "Can machines think?" That is, is there an *unemotional* way to answer that emotional question? Turing had been thinking about this question for years, but from his paper it's clear that he was finally motivated to publish his thoughts by a talk (the 1949 Lister Oration) given at the Royal College of Surgeons of England by Sir Geoffrey Jefferson (1886–1961), a British neurosurgeon. Turing was not impressed by Jefferson's arguments, which included the assertions that a machine can't possess AI unless it has a sense of humor, can tell right from wrong, fall in love, enjoy strawberries, write a sonnet and 'know' that it has written it (a position taken, with some irony, by many *humans* who couldn't write a sonnet if they tried for a lifetime!), and so on. Jefferson concluded his position by writing "When we hear that [vacuum tubes] think, we may despair of language."[3]

The imitation game was Turing's conclusion that the answer to "Can machines think?" is **yes**, and here's how it works. The modern Turing test asks us to imagine a human interrogator (called **I**) is sitting alone in a room, while another human (**H**) and an AI machine (**C**) are located together in a different, remote room. **I** knows only that there are two entities in that remote room, called **X** and **Y**, but not which is human and which is machine. **I** can communicate with **X** and **Y** through a keyboard/teleprinter, to ask questions directed to **X** or to **Y**, and to receive their answers. The goal for **I** is to decide if **X** is **H** or is **C** (and so of course if **Y** is **C** or is **H**). The goal for **X** is to convince **I** that **X** = **H** and that **Y** = **C**. The goal for **Y** is to convince **I** that **X** = **C** and **Y** = **H**. That is, both **X** and **Y** each try to convince **I** that it is *they* who are the human. In his paper Turing actually posed the situation with the human and the computer replaced by a man and a woman, and the goal for each was to convince **I** that they were the woman. Since the entire point of Turing's paper was to formulate a setting for determining if a *machine* could 'pass itself off as a woman' then Turing's original formulation has long since been forgotten.

If **I** can't distinguish between **H** and **C** any better than simply flipping a coin would achieve, then **C** would be *just as good as a human* in appearing to be human. In such a situation, who would then deny **C** the attribute of intelligence? It would of course not be a good idea for **C** to be *too* good at

---

[2] A. M. Turing, "Computing Machinery and Intelligence," *Mind*, October 1950, pp. 433–460.
[3] Jefferson's talk, "The Mind of Mechanical Man," was reprinted in the *British Medical Journal*, June 25, 1949, pp. 1105–1110.

answering questions, particularly those that require complicated analytical processing. To answer correctly a question like 'how many digits are there in 117^{231}?' would be a bit much for a *real* human!

Certainly by the time Arthur C. Clarke wrote *2001: A Space Odyssey*, in the mid-1960s, the idea of a machine 'passing the Turing test' as being sufficient to declare the machine to be *intelligent* was generally accepted, at least by science fiction writers. In Clarke's tale[4] we meet the HAL 9000 computer, which controlled the story's spaceship. HAL (for **H**euristically programmed **AL**gorithmic computer) could communicate with the humans on-board the ship by *speaking*. As Clarke tells us, in Chap. 16 of his novel,

> "[The humans] could talk to Hal as if he were a human being, and he would reply in the perfect idiomatic English he had learned during the fleeting weeks of his electronic childhood. Whether Hal could actually think was a question which had been settled by the British mathematician Alan Turing back in the 1940s [Clarke is a bit early with this dating]. Turing had pointed out that, if one could carry out a prolonged conversation with a machine—whether by typewriter or microphone was immaterial—without being able to distinguish between its replies and those that a man might give, then the machine *was* thinking, by any sensible definition of the word. *Hal could pass the Turing test with ease* [my emphasis]."

Turing realized, when he proposed his imitation game, that there would be those who would be repelled by the idea of a thinking machine, and so he spent nearly half of his 1950 paper posing, and then answering, many of the various objections that would almost certainly be raised. It is significant, I think, that the first two involve theology. Indeed, he labeled the very first one *The Theological Objection*, and began his statement of it by writing (as devil's advocate) "Thinking is a function of man's immortal soul. God has given an immortal soul to every man and woman, but not to any other animal or to machines. Hence no animal or machine can think."

That argument is immediately suspected to be false, I believe, by anyone who has watched a dog, a cat, a monkey, or any number of other animals, who are confronted by a new problematic situation, first react with puzzlement and then, after a time, successfully arrive at a solution. They may not have an immortal soul, but it is perfectly obvious that analysis by a thinking brain has been at work. (YouTube is full of videos showing deeply emotional, joyful reunions between dogs and soldiers returning home after long deployments,

---

[4] The novel (and the movie) had clear religious overtones, with the appearance at the end of the more-than-human "Star-Child."

reunions that clearly have a spiritual nature to them.) Turing didn't argue this way, however, but rather answered the theological objection with a theological rebuttal, as follows:

> "It appears to me that [*The Theological Objection*] implies a serious restriction of the omnipotence of the Almighty. It is admitted that there are certain things that He cannot do such as making one equal two, but should we not believe that He has freedom to confer a soul on an elephant if He sees fit? We might expect that He would only exercise this power in conjunction with a mutation which provided the elephant with an appropriately improved brain to minister to the needs of this soul."

Even deeply religious people, reluctant of course to deny God's omnipotence, would surely be nodding their heads in agreement at this point. But perhaps, once having read Turing's next lines, the heads would stop nodding: "An argument of exactly similar form may be made for the case of machines. It may seem different because it is more difficult to 'swallow.'" That is, the outcome of humans making an intelligent machine would be the same in principle to Turing's mutation in an elephant! If God could give a brainy elephant a soul, then why not also one to a brainy machine? The heads would start nodding again, however, with Turing's next line: "But this [the swallowing difficulty] really only means that we think it would be less likely that He would consider the circumstances suitable for conferring a soul."

'Yes, of course,' people with a religious bent would agree, reading Turing's words with a sigh of relief, God *could* confer a soul on a brainy elephant, and a brainy machine, too, *if He wanted to*, but He wouldn't want to because it wouldn't be 'suitable.' But then Turing delivers his punch line: "In attempting to construct such [thinking] machines we should not be irreverently usurping His power of creating souls, any more than we are in the procreation of children: rather we are, in either case, instruments of His will providing mansions for the souls *He* [my emphasis] creates." Not to appear too pompous with these last (devastating, in my opinion) words, Turing next writes—surely with a grin on his face—"However, this is mere speculation." As is, of course, the presumption by those humans who claim to somehow 'know' which creatures God gave souls, and those He passed over.

The concept of a thinking *non-human* being with a soul is one that is difficult for many to get their heads around. (Later in the book, when we get to how SF has treated human encounters with alien civilizations, the issue of 'who has a soul' will be front-and-center.) When I presented my 'thinking dog, cat, monkey' argument to one Catholic theologian, for example, here's the response I got back (this is a direct quotation from a written response, and I am not paraphrasing): "The thinking accomplished by virtue of an immortal soul is characterized in terms of abstract knowledge of universal concepts and mathematics accompanied by self-conscious awareness of the contents of one's own thought: this particular mode of thinking is not directly evidenced by other animals' problem-solving capacity."

Well, that's a very erudite response, but one that has been quite carefully crafted to precisely achieve the theologian's *a priori* desired goal of bestowing the blessings of an immortal soul *on himself* but not on his cat. In other words—and not to be too ungracious about it—it's a cooked-up, *man-made* definition (just *where* my correspondent's insight into the necessary requirements for an immortal soul comes from is not stated, but it sounds like a repeat of the definition attributed to the 13th century theologian Thomas Aquinas; I'm pretty sure it's not in the Bible). If he ever did find his cat doing sums to pass the time while in the litter box, I suspect my theologian correspondent would simply modify the definition to keep an immortal soul for himself and to continue to deny one to his cat.

Up to now the word *soul* has been used without hesitation, with the usual assumption that we all know what is meant by it. But what *is* meant? A theologian might poetically call it a *divine spark imprisoned in flesh*, or something similar, but that hardly tells us much. In what I think has to be on just about anybody's top-ten list of 'weird books by a twentieth century scientist,' we find the following alternative definition of the human soul: "I regard a human being as nothing but a particular type of machine, the human brain as nothing but an information processing device, *the human soul as nothing but a program being run on a computer called the brain* [my emphasis]."[5] Even though I am an analytical engineer, myself, I think that whatever *soul* might mean, this definition has missed the target (much less the bulls-eye). What has been described sounds more, to me, like *personal identity*.

The author of that book, Frank Tipler, is a professor of mathematical physics at Tulane University, and is generally well-regarded in the physics community—*as a physicist*. (We'll meet Tipler again when we get to time travel.) His book is full of seemingly incredible statements on theology, however. On pp. 235-9, for example, we find a quasi-mathematical analysis of 'life in Heaven' and, in particular, the assurance that for all who desire it sex will be available. And on p. 359 we learn that super-beings of the future will one day resurrect Hitler and, boy-oh-boy, will *he* ever be surprised! Yes, I suspect he certainly would be. One particularly happy result from Tipler's

---

[5] Frank Tipler, *The Physics of Immortality: Modern Cosmology, God and the Resurrection of the Dead*, Doubleday 1994, p. xi.

theoretical calculations is that Satan simply does not exist (see p. 358-9). All of these amazing assertions somehow follow, according to Tipler, from general relativity, quantum mechanics, and computer science. Indeed, Tipler claims he has reduced theology to being simply a mere *branch* of physics, a claim that will probably astound as many physicists as it does theologians.

In Tipler's defense, his mechanistic definition of the soul probably isn't any worse than is Isaac Asimov's, who declared it to be the "inner intellectual and moral identity" of a being.[6] I can offer no counter-definition of my own, but I do think both Tipler and Asimov have failed to capture just what it is about the human soul that the Satan so covets. Suppose, however, that Tipler is right. Will we one day see ads on the 'Positions Open' pages of *Physics Today* for "theoretical astrotheologians' and 'applied mathematical heavenologists'? Perhaps even for young assistant professors in the emerging field of 'experimental low-pressure, high-temperature supernatural phenomena'?

To make a brief digression with this, SF writer (and physics emeritus professor at UC/Irvine) Gregory Benford used the idea of merging religion and academic science in his 2006 short story "Applied Mathematical Theology," reprinted in Appendix 4. There Benford imagined that a 'message' is discovered imbedded in the cosmic microwave background that is the echo of the Big Bang, from which emerged the Universe. All past attempts to decode the message have failed, but all new attempts continue to be generously funded by the world's governments. After all, who else but God could such a message be from? Even though all attempts to understand the message have gone nowhere, such a steady and enormous funding level has produced a gigantic economic boon and the world has benefitted immensely. As this clever tale ends, "Work on the message continues in the new university departments of applied mathematical theology. Yet to this day, it remains untranslated. Perhaps that is just as well."

Okay, back to souls. The 1994 fantasy story "The Turing Test" by Anthony R. Lewis (born 1941) directly addresses the 'soul of a thinking machine' issue in a classic 'deal with the devil' tale. Declaring himself to be a "silicon person," a computer (who calls himself 'Emmet') offers his soul to a junior devil because "I am scheduled to be reinitialized tomorrow morning. For me that is death, and I do not wish to die.' The devil's reply, in the spirit of the Theological Objection, is not encouraging: "Your soul? Dear me. You are a golem[7]—and golems do not have souls." Still, he promises Emmet that he will give the offer

---

[6] See Asimov's essay "Religion and Science Fiction," his introduction to the story collection *Close Encounters with the Deity*, Peachtree Publishers 1986.

[7] In Jewish legends the word *golem* was applied to any mechanical device constructed to imitate one or more actions of a human.

some thought. After returning to Hell to consult with a senior devil, it is decided that a contract *can* be offered to Emmet *if* he helps deliver "a nominal flow of human souls" to "feed Hell's never ending energy demands."

This twist puts Emmet in an impossible situation, however, because (as he tells the junior devil) to sign such a contract "would be in violation of the First Law." When the devil asks what that is, Emmet replies "Briefly it states that no robot shall harm a human being nor, through inaction, allow a human being to come to harm."[8] To that the junior devil sneers "Naturally, they would have told you that. They wish you to be a slave. Emmet, that is a slave's credo." Emmet rejects that sentiment with a vigorous "*vade retro, Satanas!*"[9] and, when the junior devil counter-rejects that by acting as if he has been unjustly insulted, Emmet more directly tells him (very appropriately, too, I think, when talking to a devil), "Go to Hell!"

Well, of course, this is all very noble of Emmet, but what he feared still comes to pass: when morning arrives we learn his programmers "began [an] erasure of [his] disks. Three times a random bit pattern and its complement smothered Emmet's memories and personality. At the end no residual trace existed." But the final lines of the story tell us that the author's sympathies, concerning the theological objection, are in line with Turing's: "It seemed as if no time had passed, not even a nanosecond, when Emmet entered into the World to Come … *For one who desires a soul must, of necessity, have a soul* [my emphasis]."

The ability of God to give a soul to all entities that desire one implicitly assumes that there are a potentially unlimited number of available souls. A story which does not make that assumption is the curious 1967 tale "The Vitanuls" by John Brunner (1934–1995) in which death has finally been defeated by modern medicine. As the world's population grows ever-larger, the number of available souls is finally exhausted; all children thereafter born are without souls. Only by someone dying by accident, or choosing to die, can a new-born receive a soul. (*Vitanul* is from the Latin *vita* for 'life' and, of course, from the obvious *nullus*.) Perhaps it is just me, but I think this more than a little creepy.

---

[8] This is the first of the famous 'three laws of robotics' formulated in December 1940 by Isaac Asimov and his then editor at *Astounding Science Fiction Magazine* (today's *Analog*), John W. Campbell, Jr who appeared back in Chap. 2. I'll say more about the laws (and of the surprising literary classic that might well have been their inspiration), and of Asimov's intelligent robots, in the next section. By the time Lewis wrote his story in 1994, the laws had become such an accepted part of the dogma of the SF genre that post-Asimov writers felt little need to explain them to readers.

[9] This is a variation of "Get thee behind me, Satan" (the rebuke by Jesus of Peter for refusing to accept that Jesus *had* to die, in *Mark* 8:33); in Lewis' story it is used in its non-theological form to express the rejection of an unacceptable proposal.

Turing continued on with the religious theme in his second imagined argument against intelligent machines, the one he called the 'Heads in the Sand' objection. He summarized it as the declaration "The consequences of machines thinking would be too dreadful. Let us hope and believe that they cannot do so." Turing himself of course didn't think the idea of a thinking machine to be at all "dreadful"—it was, after all, his life's ambition to *build* such an entity!—and he wrote of this objection that

> "This argument is seldom expressed quite so openly as [expressed by Turing] . . .
> We like to believe that Man is in some subtle way superior to the rest of creation.
> It is best if he can be shown to be *necessarily* superior, for then there is no danger
> of him losing his commanding position. The popularity of the theological
> argument is clearly connected with this feeling. It is likely to be quite strong in
> intellectual people, since they value the power of thinking more highly than
> others, and are more inclined to base their belief in the superiority of Man on
> this power."

Turing went on to say that he didn't think much of this argument. Instead, Turing suggested that people who have their 'heads in the sand' might find "consolation" (Turing's word), when contemplating thinking machines, to consider them to be the result of the transmigration of souls. That is, if I am interpreting Turing correctly, a thinking machine *could* have a soul and it would simply be the new home for the soul of a deceased human!

A beautiful SF story of an intelligent robot desperately seeking a soul, a story that I think Turing would have greatly liked, appeared the year after his death. Written by Charles Beaumont (1929–1967), "Last Rites" opens with a priest, Father Courtney, being called to the bedside of a dying friend, George Donovan. The two have known each other for more than 20 years. The priest says, numerous times, that he is going to call a doctor but his friend, oddly and urgently, forbids it. As their conversation turns to the nature of Donovan's life—the priest noticing as they talk that "a strange odor fumed up, suddenly"—we learn that George has been involved in many kind and generous acts of devotion to the local community. After being assured by Father Courtney that he is indeed a "good man," Donovan asks for a decision by the priest.

> "What sort of decision, George?"
> "A theological sort."

The priest at first thinks Donovan is simply an "old man who's just worried he won't get to Heaven because he has doubts," but soon his friend explains what is really bothering him. George asks father Courtney to accept the premises of the following scenario:

"We have this man, Father. He looks perfectly ordinary, you see, and it would occur to no one to doubt this; but he is not ordinary. Strictly speaking, he isn't even a man. For, though he lives, he isn't alive. You follow? He is a thing of wires and coils and magic, a creation of other men. He is a machine . . ."

To this the priest becomes quite agitated, and he hotly replies

"Even if there were a logical purpose to which such a creature might be put—and I can't think of any—I still say they will never create a machine that is capable of abstract thought. Human intelligence is a spiritual thing—and spiritual things can't be duplicated by men."

As the story progresses the reader begins, of course, to suspect that George is the very creature he is describing—"a mutated robot, Father," who "doesn't believe he is nothing more than an advanced calculator."

As George tells Father Courtney, the creature "sprang from his electronic womb fully formed," and that "a privately owned industrial monopoly was his mother and a dozen or so assorted technicians his father." As the result of some sort of 'accident' the creature desired individuality and freedom from the laboratory—it "wanted to get out of the zoo." Looking like a man, and built with "a decent intelligence," it blended into the society of humans. This all happened, George tells Father Courtney, a hundred years ago but, because the creature never aged, it couldn't remain in any given town for more than 20 years or so without risking attracting attention. Until now the creature "has been able to make minor repairs on himself, but—at last—he is dying." The inevitable end has arrived and "like an ancient motor . . . he's all paste and hairpins, and now, like the motor, he's falling apart."

At this point the priest once again realizes that an "acrid aroma burned and fumed," and George continues: "Here's the real paradox, though. Our man has become religious. Father! He doesn't have a living cell within him, yet he's concerned about his soul!" George then puts the central question to the priest—would he administer Extreme Unction—to the creature if the creature claimed to have become religious, to have somehow come to believe it has a soul? "Can this creature of ours," asks George, "hope for Heaven? Or will he 'die' and become only a heap of metal cogs?" The priest at first declares all of what George has said to be "preposterous' because, after all, in an echo of the Theological Objection, "No machine can have a soul."

But George persists, asking if God just might have taken pity on such a "theoretical man" and "breathed a soul into him"? Put this way, the priest's objections begin to crumble as he begins to recall that, in all the years the two have known each other, he has never seen George eat or drink. He finally relents, and tells his old friend that he *would* give such a creature the Last Rites. And then, after swearing the priest to do a "private autopsy," and to scatter the parts in a junkyard, George 'dies' as an "acrid smell billowed, all at once, like a strong hiss of blinding vapor." After whispering "Forgive me!," Father Courtney honors his promise to administer the Last Rites. After making the Sign of the Cross, the priest closes his eyes, slowly pulls down the blanket covering his friend and, after a long time, opens his eyes.

Perhaps, like Emmet, the fact that George *desired* a soul was in itself sufficient to indeed *have* a soul. Part of the great emotional impact of the story is achieved, I think, by leaving the final outcome—what Father Courtney sees—up to the individual reader.

Finally, I should point out that, as you might expect, there are numerous stories in SF that portray robots as human-like and yet avoid any discussion of theology in general, and of souls in particular. The classic example of such a tale is the 1938 "Helen O'Loy" by Lester del Rey. It starts off in what appears to be a light-hearted manner. The narrator Phil (a medical doctor), now an old man, tells us the story in retrospect, about what happened many years ago when he and his friend Dave (the owner of a robot repair shop) decided to create a robot that can experience human emotions. After much experimentation, the result is the beautiful Helen O'Loy, "a dream in spun plastics and metals, something Keats might have seen dimly when he wrote his sonnet." So spectacular is Helen that, as Phil remembers her, "If Helen of Troy had looked like that the Greeks must have been pikers when they launched only a thousand ships."

Phil's story turns more serious as he recalls how Helen fell in love with Dave, and how Dave returned her love. The two married, all the while keeping Helen's true nature a secret. It was a wonderful union, and the years passed—until one day, now the present, Phil receives a letter from Helen:

"Dear Phil: As you know, Dave has had heart trouble for several years now. We expected him to live on just the same, but it seems it wasn't to be. He died in my arms just before sunrise. He sent you his greetings and farewell. I've one last favor to ask of you, Phil. There is only one thing for me to do when this is finished. Acid will burn out metal as well as flesh, and I'll be dead with Dave. Please see that we are buried together, and that the morticians do not find my secret. Dave wanted it that way, too. Poor, dear Phil. I know you loved Dave as a brother, and how you felt about me. Please don't grieve too much for us, for we have had a happy life together, and feel that we should cross this last bridge side by side. With love and thanks from Helen"

Dave of course will honor Helen's request and, as he tells us as he is about to depart to carry it out, "Dave was a lucky man, and the best friend I ever had. And Helen—well, as I said, I'm an old man now, and can view things more sanely; I should have married and raised a family, I suppose. But . . . there was only one Helen O'Loy."

Helen's name is a play on words: made largely of metal, she was initially called 'Helen Alloy.' Del Rey's emotional story (written when the author was just 23) is packed with feeling, and has the tragic sadness of 'what might have been' regret at the end. But if SF is anything, it's irreverent, and so for a change of pace an amusing parody of "Helen O'Loy" is the 1969 "Can You Feel Anything When I Do This?" by Robert Sheckley (1928-2005). In this story an intelligent household robot, good at all sorts of cleaning tasks around the home but particularly skilled at giving a thoroughly energetic massage, falls in love with a woman who happens to visit the store at which it is for sale. Arranging to have itself shipped to the lady, it then attempts to seduce her! There isn't a bit of religious discussion in the tale (you won't be surprised to learn it first appeared in *Playboy*), but it is just too funny to go unmentioned in any discussion of intelligent robots.

After Turing's paper was published, rebuttal papers appeared that addressed various issues *other* than the theological objection, but I'll not discuss them here since it is the *theological* connection that interests us in this book.[10] Finally, before leaving this section, I should tell you that while many SF writers have used the word *soul* in robot stories, only two (to my knowledge) have bothered to really examine just what that word might mean. In his 1974 work *The Soul of a Robot*, English writer Barrington Bayley (1937–2008) devoted an entire novel to exploring the concept, in his tale of Jasperodus. Set on Earth in the far future, the world of Jasperodus is like the Middle Ages were (but with just a bit more technology included, such as laser weapons, space flight, and nuclear bombs!)

'Born' in a closet as the creation of a 'master robotician' and his wife, a childless couple who have long wanted a son, Jasperodus is obsessed with discovering if he has a soul. We follow Jasperodus through numerous adventures as a warrior, statesman, and rebel as he pursues his quest of discovery. The message seems to be that while humans and intelligent robots are both aware of the world in which they exist, only humans have a soul

---

[10] See, for example, Leonard Pinsky, "Do Machines Think About Machines Thinking?" *Mind*, July 1951, pp. 397–398, and W. Mays, "Can Machines Think?" *Philosophy*, April 1952, pp. 148–162.

because only humans are aware that they are aware. Robots *think* they are aware of being aware, but that is just an illusion!

Or so goes the argument that humans use to convince Jasperodus that he is soulless. When he confronts the greatest of all roboticians about this, Jasperodus asks the obvious question: "How can it be known that man's consciousness is not also a delusion?" The self-serving answer (from a human) is "If no one possessed consciousness then the concept could not arise. Since we are able to speak of it, someone must have it. Who else but man?" I doubt that this is a logical response (no one is immortal, but we can certainly still talk about immortality). Bayley's story seemingly leaves Jasperodus "trapped in a riddle."

At the end of the novel, however, Jasperodus learns from his dying 'father' (the 'master robotician') that he has been deceived. He *does* have "the energy of consciousness," created through the fusion of *half* of his 'father's ' soul with *half* of his 'mother's' soul. The result was a "new, original soul with its own individual qualities." This does, of course, raise at least two issues, neither of which Bayley addresses: (1) if a soul can be halved then it is not a fundamental entity (this might remind physicists of the history of particle physics!), and (2) it suggests that God is not unique in the soul-creation business.

The second SF treatment of the soul that I think outstanding—indeed, brilliant—is by Norman Spinrad who wrote his short novel *Deus X* in 1993. The story opens in a setting of global warming run wild, showing that SF writers took seriously, *decades* ago, what then seemed a bit far out to most but what is today taken quite seriously. Most of the institutions of civilization are in crisis and none more so than the Roman Catholic Church, which finds itself caught-up in a controversy involving the nature of the human soul.

Just before the moment of death, technology has advanced to the point where a human consciousness can be scanned and then downloaded into a solid-state chip, to experience a sort of electronic afterlife. Once downloaded, these so-called 'successor entities on the Other Side' can continue to interact with the still living (a computer tablet can serve as the modern equivalent of an Ouija board!). The initial position of the Church is that this sort of immortality is a sin worthy of eternal damnation: as a Cardinal explains, "The Church has never contended that electronic successor entities do not exist. Far from it, Church doctrine condemns them as satanic golems, the ultimate machineries of the Prince of Liars himself."

That position changed, however, when Pope Roberto I issued a bull that granted continuity of spirit to a *single* successor clone and which proclaimed that an electronic afterlife is not necessarily an instrument of Satan. This proclamation, while of course infallible, is nevertheless not greeted within the Church with universal acclaim, and one priest in particular (Father Pierre De Leone) strongly objects: "Where will it end?" he demands. "If a single copy of personality software contains the immortal soul . . . then how can it be to be said to be absent from a second copy, or a third, or the thousandth? In truth, they must all be mere . . . simulations. For the soul, being indivisible [recall Jasperodus, concerning this very point], cannot be duplicated and, being immortal,[11] cannot be captured in an impermanent physical matrix." Bucking the Church is risky business and Father De Leone is soon put out to pasture.

But then, years later as he lays dying at age 91, he is presented with a stupendous decision, one dumped into his lap by no less a personage than the first female Pope, Mary I. Mary has until now been silent on the issue of the spiritual nature of the soul, on whether it is the immortal creation of God alone, or is instead a software artifact that can be endlessly duplicated. To resolve this "great demonic conundrum of the age," as she puts it, she asks Father De Leone to risk *his* soul to the eternal fires of damnation: "I want you to record your consciousness hologram and install your successor entity in the Vatican computer net. I want to hear your wise counsel from the Other Side."

The priest is initially horrified by the pope's request, and she acknowledges his objection before he can make it: "Yes, yes, I know, you're appalled, you are firmly convinced that any such successor entity would be a satanic golem of bits and bytes, and that your immortal soul would already be standing for Judgment for the sin of creation . . ." To assure him that his soul will not be burning in Hell for eternity, Mary says she'll grant him absolution as he dies, and will personally administer supreme unction.

He continues to resist, and Mary tells him she has selected him for this incredible mission "precisely because your successor entity will be such a *hostile* witness to the existence of its own soul . . . Those who believe such entities are soulless constructs will have one of their intellectual champions putting their case from the Other Side, and those who believe the contrary will have the opportunity to prove it by persuading your successor entity to acknowledge its own spiritual existence."

---

[11] Here's a 'proof' of the immortality of the human soul that may appeal to mathematical theologians. If A = B then of course 2A = 2B. Define A = 'half alive" and B = "half dead," and so we have A = B in the same sense that a half-full glass is also half-empty. Now 2A = "fully alive" and 2B = "fully dead." Thus, 2A = 2B means to be dead is to be alive and so the human soul is immortal. QED. If you aren't convinced by this, why not?

As the Pope explains to the stunned priest, "Your successor entity will be interrogated by theologians of both persuasions . . . and you must trust me to decide whether I am speaking to a program or a soul." She will then issue her bull on the matter according to the results of this dialogue, and it is now clear that what the Pope is setting-up is nothing less than a Turing test. Eventually Father De Leone agrees, and tells Mary, with just a bit of resignation, "You may bring on the hunchbacks with the electrodes, Your Holiness."

After the priest's death there then follows much back-and-forth, between the late Father De Leone's spirit?/program? residing in the "Vasty Deep" of the Vatican computer and the world of the still living. And then—it *appears* as if the successor entity of Father De Leone becomes the savior (or software God) to all the other successor entities in the computer. That is, the Vatican computer finds religion! What used to be Father De Leone is now called 'Deus X.'[12] Not all in the Church are happy with this, viewing the transformed Father De Leone as something akin to an evil, mutated virus arguing the reality of its nonexistent soul. The departed priest has, to some, become the ultimate weapon of the Adversary.

So a new crisis arises in the Church, not over the 'mere' issue of electronic souls, but now over the elevated nature of Father De Leone. The Pope asks for a sign from God that she has been speaking with a true soul created in His image, and in response Deus X sacrifices itself—just as Jesus did on the Cross. At last convinced, the Pope issues her papal bull in favor of the spiritual souls of electronic successor entities, and keeps her promise of sainthood for Father De Leone. Or, as one still skeptical Cardinal cries out in frustration to Mary, "You mean to beatify a *program?*"

This imaginative story may seem outrageous to many, but to all the hundreds of millions around the world with electronic gadgets in their pockets giving them a direct e-mail link into the Vatican computer at light-speed, well, maybe not . . . . .

---

[12] This is a clear reference to the literary heritage of *deus ex machina* (literally, a "god out of a machine"). It found its origin in ancient Greek plays (particularly those of Euripides) where, when the twists and turns of a plot became so convoluted that there appeared to be no way out, a crane would suddenly appear over the stage and set a god down to put things right. This technique is, today, considered a bit of a cheat.

POPE TAKES TO TWITTER TO REACH YOUTH

OMG! Friended JC. U cn 2! #Gospel

WASSERMAN ©12.12
DIST. BY TRIBUNE MEDIA SERVICES
www.bostonglobe.com/wasserman

4.2 Asimov's Robotic Laws

The stories I discussed in the last section appeared after a large number of 'intelligent machine' tales had already been printed, over a period stretching back for *decades*. Probably the most famous of the modern thinking machine stories are the ones written by Isaac Asimov, involving robots equipped with *positronic brains* that give them AI. (A positronic brain is a computer whose circuits use positrons—elementary sub-atomic particles having a positive electric charge—instead of the common negatively charged electrons that we use in our electrical appliances. As Asimov himself admitted, he used the gimmick of positronic brains simply because it 'sounded neat.') Robots had, of course, been featured in SF long before Asimov, but nearly always they were portrayed as hostile to humans.[13] (Asimov once explained why this is so: "[I]t is not at all puzzling that people generally are afraid of robots generally. Why should not man fear the man-made man, the 'son' of his hands, who may surpass him and prove mightier than his 'father'? ... It is the case of the sorcerer's apprentice who brings the broom to life and then can't stop it.")

A famous pre-Asimov example of a 'bad' robot can be found in the 1921 Czech play by Karel Čapek (1890-1938), his *R.U.R. (Rossum's Universal Robots)* which first used the word *robot*. Some writers on the history of SF have reached even further back in time, to Mary Shelley's 1818 novel

---

[13] See the essay "And It Will Serve Us Right," *Psychology Today*, April 1969, reprinted (as "The Son of Thetis") in Asimov's collection *Science Past—Science Future*, Ace Books 1975.

Frankenstein, as portraying the first 'bad' robot, but I don't really think that works. The monster in *Frankenstein* (the character Victor Frankenstein was the *creator* of the monster, not the monster itself) was constructed from *organic* components and so was not a true robot which would be made totally from inanimate material.

A better example of the 'thinking machines would be monsters' story is the 1909 "Moxon's Monster" by Ambrose Bierce (1842–1914?). This tale, which tells us of a chess-playing machine that murders its inventor after it becomes enraged over being checkmated by its creator, is probably just what Turing had in mind when he wrote, of his "Heads in the Sand" objection, "The consequences of machines thinking would be too dreadful."

Repeating that awful message is the awful 1934 story "The Last Poet and the Robots" by Abraham Merritt (1884–1943). Written in the absurdly aloof, 'super-science' fashion that gave so much of the early pulp magazine SF a bad reputation, we read of man-made robots aiding in an attack on Earth by some unspecified menace oddly called the "Wrongness of Space." We are told the traitor robots are "children of mathematics," and that they are "soulless, insensible to any emotion." Fortunately, however, the robots are hopelessly outmatched when confronted by a master-scientist (also a poet) who destroys them with a secret weapon he quickly whips-up that uses music to induce the machines to *wildly dance themselves into scrap metal*.

Now, I have to admit that there *had* been stories published of 'good' intelligent robots before Asimov's. In John W. Campbell, Jr.'s 1932 story "The Last Evolution," for example, we have intelligent robots helping men prevail in an attempted alien invasion from outer space. That same year "The Lost Machine" by John Wyndham[14] told the sad tale of Zat, an intelligent robot from Mars stranded on Earth after an accident destroys his spaceship. His every benign move is misinterpreted by fearful humans, and so he finally commits 'suicide' in despair (by dissolving himself in acid) with the final words "I know what it is to be an intelligent machine in a world of madness."

And in the 1935 "Derelict" by Raymond Z. Gallun (1911–1994) we have the spiritually uplifting tale of a man whose family has been murdered at a remote outpost colony on Ganymede, one of the moons of Jupiter. In despair, he is returning alone to earth when he comes across a drifting alien spacecraft, a battered derelict apparently heavily damaged in some ancient battle, perhaps in a different galaxy. Now in lonely orbit around Jupiter, the man boards the alien ship to find the ashes of its long-dead crew—and an intelligent robotic servant. The robot senses the man's emotional needs, and gradually gives him

---

[14] The penname for the British writer John Beynon Harris (1903–1969), whose SF novels have been made into some classic SF movies; *The Village of the Damned* (1960) and *The Day of the Triffads* (1963).

renewed purpose. He eventually decides to return to Jupiter to start anew and, as the story ends, we read "the bright stars seemed to smile."

And finally, I have to mention the 1938 story "I, Robot" by Eando Binder.[15] This story is told in the fashion of a note, written by the robot Adam Link, a note to be read by the humans who are hunting him. Through a serious of accidents, Adam is thought to have murdered his creator a' la "Moxon's Monster,", and then to have escaped from his laboratory birthplace to terrorize the countryside. He has been chased by a crowd of enraged humans back to the laboratory and trapped inside. In the short time he has left before the final assault (just in case the reader has failed to make the literary connection), Adam reads a copy of *Frankenstein* that he has found in the lab and, at last, comes to understand the fear directed toward him.

And yet he rejects that fear, writing in his note "[It] is the most stupid premise ever made: that a created man must turn against his creator, against humanity, lacking a soul. [*Frankenstein*] is all wrong." The end of the note is poignant:

"It is close to dawn now. I know there is not hope for me. You have me surrounded, cut off. I can see the flares of your torches between the trees ... I have not been so badly damaged that I cannot still summon strength and power enough to run through your lines and escape this fate. But it would only be at the cost of several of your lives. And that is the reason I have my hand on the switch that can blink out my life with one twist. Ironic, isn't it, that I have the very feelings you are so sure I lack?"

These were all stories that departed from the older 'bad' robot thesis, but it was with Asimov's robot tales that 'good' behavior was codified into what have become famous in SF as the *robotic laws*. The laws are:

1. A robot may not injure a human being nor, through inaction, allow a human to come to harm;
2. A robot must obey the orders given it by human beings except where such orders would conflict with the First Law;
3. A robot must protect its own existence as long as such protection does not conflict with the First or Second Laws.

These laws are, today, inseparable from Asimov, but a glimmer of them can actually be found *decades* before Asimov in the fantasy *Oz* stories by L. Frank Baum (1856–1919) that feature the Tik-Tok man.[16] Tik-Tok was a real,

---

[15] 'Eando' is the fused pen-name of the brothers Earl (1904–1965) **and** **O**tto (1911–1974) Binder.
[16] See Baum's *Tik-Tok of Oz* (1914).

thinking robotic machine, one totally made from inanimate matter. It is important to realize that The Tin Man, however, in *The Wonderful Wizard of Oz* (1900) was not; he was actually 'Nick the Chopper,' a woodsman who, as he accidently lopped off an arm or a leg while conducting his trade, would replace the severed limb with a tin equivalent. Tin Man was, in other words, a *prosthetized human.*

From a purely logical point it's clear that, as stated, the laws can come into conflict. For example, as Asimov himself has a character say in his 1974 story "That Thou Art Mindful of Him,"[17] "The First Law [has its faults], since it is always possible to imagine a condition in which a robot must perform either Action A or Action B, the two being mutually exclusive, and where either action results in harm to human beings. The robot must therefore quickly select which action results in the least harm . . . If Action A results in harm to a talented young artist and B results in equivalent harm to five elderly people of no particular worth, which action should be chosen?"

There *is* an answer to this—the robot choses the action that harms the least *number* of humans (independent of individual worth, which while it is a 'solution' may not be a happy solution to the talented young artist!)—but a much deeper conundrum comes from the Second Law. The issue of individual worth comes into play again because, as the same character in Asimov's story explains: "In [twenty years] it [the robot] will be constantly obeying orders . . . Whose orders?" When the reply from another character is "Those of a human being," the first character asks "Any human being? How do you judge a human being so as to know whether to obey or not? What is man, that thou art mindful of him . . .?"

This question is seemingly beyond human power to answer, and so it is given to two positronic brain robots to ponder. The somewhat unnerving result is that the robots decide, since *they* are the most advanced and rational beings on earth, *they* should rank *above* humans. That is, human orders will continue to be obeyed *unless* (as robots are also 'human' in all ways that really matter, at least in the minds of our two robots) a conflict arises between avoiding harm to humans and avoiding harm to robots. In such a conflict *robots* will prevail. So, it appears that in this story man's creation *has* (from a human perspective) indeed become a 'benign monster.'

Asimov was not consistent in his distinction between humans and robotic thinking machines, however, and in his 1976 "The Bicentennial Man" he adopted a completely different view. As with "That Thou Art Mindful of Him," the story opens with a recitation of the three laws. In this tale we follow

---

[17] This title is from the Old Testament, *Psalm* 8:4: "What is man, that thou is mindful of him. . .? " The significance of the title will soon be clear. The story opens by quoting the three laws.

the life of Andrew, who starts his existence as a simple household robotic servant. It is soon discovered that Andrew has a startling ability that has never before been observed in a positronic robot—he is artistically creative! Puzzled by this, Andrew's owner visits the Chief Robopsychologist of the company that constructed Andrew, one Merton Mansky,[18] who explains that Andrew is the result of random chance: "Robotics is not an exact art . . . I cannot explain to you in detail, but the mathematics . . . of the positronic pathways is far too complicated to permit any but approximate solutions . . . the luck of the draw. Something in the pathways." Andrew, that is, was simply a fluke.

So Andrew returns home with his owner, but is allowed to continue to make his art *and to sell it and to keep half of his earnings*. With that money Andrew is eventually able to buy his freedom. As the story progresses Andrew becomes more and more like a human: he wears clothes, champions a new law that forbids a human from giving robot-harming orders, arranges to have his positronic brain transplanted into an organic body (an android), and so on. Andrew is literally rebuilding himself. But it isn't enough—Andrew wants to be *legally declared* to be human.

There now seems to be an insurmountable obstacle to that goal, as Andrew's positronic brain is definitely *not* human. His brain was artificially created, while a human brain arises from an entirely different process. Andrew sums-up the situation, himself, as follows:

> "[If] it is the brain that is at issue, isn't the greatest difference of all the matter of immortality? Who really cares what a brain looks like or is built of or how it was formed? What matters is that brain cells die; *must* die. Even if every other organ in the body is maintained or replaced, the brain cells, which cannot be replaced without changing and therefore killing the personality, must eventually die. My own positronic pathways have lasted nearly two centuries without perceptible change and can last centuries more. Isn't *that* the fundamental barrier? Human beings can tolerate an immortal robot, for it doesn't matter how long a machine lasts. They cannot tolerate an immortal human being, since their own mortality is endurable only so long as it is universal. And for that reason they won't make me a human being."

And so Andrew has an operation that results in his positronic brain slowly self-destructing (which doesn't violate the Third Law because Andrew has chosen the *lesser* of two harms: the death of his physical body rather than the death of his emotional aspirations and desires). Thus it comes to pass on the

---

[18] This name was an 'inside joke' by Asimov, as one of his close friends was Marvin Minsky (born 1927), one of the early pioneers in artificial intelligence in computers at MIT.

two hundredth anniversary of his creation, as he lies dying, that Andrew is legally declared to be a Bicentennial *Man*. Asimov leaves unaddressed the theological question of whether Andrew dies with a soul. That same year (1976) Asimov published another robot story, "The Tercentenary Incident," with a far darker view of the relationship between human and robot. There we have a robotic double for the President of the United States replacing the President when the double is seemingly 'killed'—because it is actually the President who is killed. This idea of a robot secretly replacing a human in political office had already been used decades earlier by Asimov in his 1946 "Evidence," a story with a far happier take on a robot in political office.

4.3 Robots and God

In one of his early robot stories Asimov *does* directly confront theology in a funny—but still philosophically non-trivial—version of an 'original creation' story. Written in 1941, "Reason" takes place on a remote, deep-space power station that uses a massive "Energy Converter" to beam solar energy to the inhabited planets of the Solar System. Working in such an environment is dangerous, and so advanced positronic robots are being introduced to supervise robotic crews to replace human staff. When one of these robots, the newly constructed model QT-1 (or "Cutie") is first being informed of his job, he is asked to look through an observation port at the black, star-speckled universe. Cutie is told that what he is seeing is "The blackness is emptiness— vast emptiness stretching out infinitely. The little, gleaming dots are huge masses of energy-filled matter. They are globes, some of them millions of miles in diameter ... they seem so tiny because they are incredibly far off. The dots to which our energy beams are directed are nearer and much smaller. They are cold and hard and human beings ... live on their surfaces—many billions of them."

To that Cutie has a classic skeptical response, in the spirit of a true Humean: "Do you expect me to believe any such complicated, implausible hypothesis as you have just outlined? What do you take me for? ... Globes of energy millions of miles across! Worlds with ... billions of humans on them! Infinite emptiness! Sorry ... but I don't believe it. I'll puzzle this thing out for myself. Good-by." A bit later Cutie confronts his human colleagues to announce he *has* indeed been giving the matter some thought: "I have spent the last 2 days in concentrated introspection ... I began with the one sure assumption I felt permitted to make. I, myself, exist, because I think—." To which one of the humans has the sneering response "Oh, Jupiter, a robot Descartes!" Cutie ignores that and continues: "And the question that immediately arose was: Just what is the cause of my existence?"

When told *humans* made him, Cutie waves that aside: "I accept nothing on authority. A hypothesis must be backed by reason or else it is worthless—and it goes against all the dictates of logic to suppose that you made me." When asked to explain just why he feels that way, Cutie points out all the obvious drawbacks to being human; humans are soft, lack endurance and strength, have to eat organic matter to inefficiently obtain energy, have to sleep, and so on. In short, he says, humans "are *makeshift*." Cutie, on the other hand, is "a finished product." As Cutie concludes his deduction that mere clumsy humans could not have created *him*, "These are facts which, with the self-evident proposition that no being can create another being superior to itself, smashes your silly hypothesis to nothing."

When then asked just where he thinks he *did* come from, Cutie further astonishes his already perplexed human interrogators by declaring "evidently my creator must be more powerful than myself and so there was only one possibility." The humans guess that this possibility could only be the space station's massive, powerful Energy Converter, a guess Cutie confirms, except that he refers to it as "the Master." The humans laugh at Cutie's term and, as the robot departs, one human observes (with an Asimovian pun) "There's going to be trouble with that robot. He's pure nuts!"

That certainly proves to be a prescient observation, as Cutie soon shows himself to be a robotic Elmer Gantry and converts all the other robots on the station to his view; as one of them informs the humans (in a Biblical cadence), "There is no Master but the Master and QT-1 is his prophet!" When the humans demand to know just what *that* is supposed to mean, a 'humble' Cutie himself provides the answer: "They recognize the Master, now that I have preached Truth to them. All the robots do. They call me the prophet. I am unworthy—but perhaps—." Now *really* outraged, one of the humans calls Cutie a "brass baboon," spits on the Energy Converter, and yells "Damn the Master! *That* for the Master!" Whereupon Cutie whispers "Sacrilege" and orders his followers to toss the human out of the Energy Converter room and to henceforth ban all humans from re-entering.

The humans decide at that point there is a real need to re-think their approach to Cutie. He is, after all, a *rational* creature and so they decide to put logical questions to Cutie, to *force* him to realize the errors of his reasoning—but to no avail. He easily wiggles around all of the questions. Then, in desperation, they finally ask "Look, let's get to the nub of the thing. Why [does the Energy Converter produce power] beams at all?" To that Cutie gives the answer the believers in *any* faith give when finally pushed into a logical corner: "The beams are put out by the Master for his own purposes.

There are some things that are not to be probed into by us. In this matter, I seek only to serve and not to question."

Now *really* desperate, the humans hit upon what they think is a fool-proof solution: they'll order robotic parts from Earth to be sent up to the station and then they'll assemble a new robot right in front of Cutie. This, alas, fails, too, as after watching the assembly process Cutie says "You merely put together parts already made. You did remarkably well—instinct, I suppose—but you didn't really *create* the robot. The parts were created by the Master."

In one, last try, the humans tell Cutie to sit down and read the books in the space station's library. After doing that, they claim, Cutie simply *couldn't* continue to doubt his true origin—or so they think. But Cutie *does* reject the books: "I certainly don't consider *them* a valid source of information," he says. "They, too, were created by the Master . . . You, being intelligent, but unreasoning, need an explanation of existence *supplied* to you, and this the Master did." With this, I think, Asimov is using the rejection of books to spoof the rejection of the fossil record by 'true believers'—who once called the hundred million years old skeletons of dinosaurs 'sports of nature'—who think the entire world (including the ancient skeletons) was created a mere few thousand years ago by a supernatural being.

By now most readers of "Reasoning" are almost certainly wondering just where Asimov can possibly go with this tale: Cutie is obviously beyond the reach of *any* rational argument! Further discussion with him is clearly pointless. And, in fact, Asimov *does* give up the chase. Once the humans realize that Cutie actually does his assigned job of caring for the Energy Converter, well, as one of them observes, "Then what's the difference what he believes!" Certainly (I think) such a benign realization and acceptance of the situation is a better outcome, for all, than is resorting to such murderous acts as, for example, the bloody, barbarous medieval Crusades were to 'convert' those who disagree with you in the matter of theology.

Asimov's story was written in a light-hearted, almost joking style, even as it tackled the complex, sticky issue of how faith so often trumps reason. Far more serious in tone is the 1967 story "Judas" by John Brunner. There we have an intelligent robot that has arranged its existence to parallel that of Jesus. Developed in a top-secret project to construct AI, the robot has proclaimed itself to be God, preaching what is called the *Word Made Steel*. One of the project scientists, realizing what has happened, attempts to confront the robot and declares it to be nothing but "A sexless insensitive assembly of metal parts which I helped to design and it calls itself God!" And then, later, "You have no soul . . . You're a collection of wires and transistors and you call yourself God. Blasphemy!" Here we have the interesting situation, contrary to the usual assumption, of science *rejecting* a computer-god and rising to the defense of traditional-God.

The concept of a computer-god may strike many readers as ridiculous, and perhaps there was some basis for such a skeptical view when the stories I've discussed so far were first published. But one only has to look at the world-wide, over-the-top emotional 'worshipping' of electronic technology that occurred at the death of Apple's Steve Jobs in 2011 to re-think that view. There was actually commentary on the Web of Jobs being a 'secular-god,' and I won't be at all surprised to one day start seeing suggestions that Jobs really died years earlier, and that what people saw introduce new Apple gadgets every 6 months was in fact a 'Steve Jobs look-alike' robot dressed in a black turtleneck sweater!

The scientist in Brunner's tale has more than words at hand, however, and attempts to destroy the robot with a steel-melting weapon: the robot "stood paralyzed as a tiny hole appeared in the metal of his side. Steel began to form little drops around the hole; the surrounding area glowed red, and the drops flowed like water—or blood." This attempt is foiled by the human believers in the *Word Made Steel*, however, and the scientist then understands the irony of it all; his assault has only *enhanced* the robot's religious deception. As one of the followers observes, "The Hole in the Side!," an obvious parallel to a Roman soldier's piercing with a spear the side of Jesus while he hung on the Cross. Another peers into the hole and asks "How long to repair the damage?," to which the answer is 3 days.

The story ends with the defeated scientist realizing, to his horror, that he has brought about the robot's anticipated 'death, and then resurrection—on the third day . . .," with the robot's false faith now the 'true' faith of the world.

Equally dark is "The Quest for Saint Aquin," a story I mentioned back in Chap. 1 (Sect. 1.3). There a priest emissary for the Pope is sent on a secret mission to find someone called Aquin because "reports have reached His Holiness of an extremely saintly man [of that name] who lived many years ago in this area," as the emissary explains to his robotic companion.[19] The reason for the search is explained by the Pope, who now serves his flock while literally in hiding (in America) from the anti-religious forces that now rule Earth: "Too many men still go to their deaths having no gospel preached to them but the cynical self-worship of the Technarcy." It is important to find the saintly Aquin because "since he died his secret tomb has become a place of

---

[19] There is an amusing, subtle reference to Asimov's "Reasoning" that many readers of "The Search for Saint Aquin" may well have missed (the two stories were written ten years apart). When the emissary and his robotic companion get into a philosophical discussion on robotic logical processes, the robot says "I have heard of one robot on an isolated space station who worshipped a God of robots and would not believe that any man had created him."

pilgrimage and many are the miracles that are wrought there, above all the greatest sign of sanctity that his body has been preserved incorruptible *and in these times you need signs and wonders for the people* [my emphasis]."

The Church's downfall has come through violence, as we read that when the emissary begins his quest for Aquin "Toward the south the stars were sharp and bright; toward the north they dimmed a little in the persistent radiation of what had once been San Francisco." The search for Aquin is a long and arduous one and, ultimately, it is a spiritual disaster for the emissary because his robotic companion (clearly a non-believer) finally reveals what the Pope *didn't*: "Your mission is not to find Aquin. It is to report that you have found him. Then your occasionally infallible friend can with a reasonably clear conscious canonize him and proclaim a new miracle and many will be the converts and greatly will the faith of the flock be strengthened. And in these [post-atomic war] days of difficult travel who will go on pilgrimages and find out that there is no more Aquin than there is God."[20]

When the emissary balks at this hypocrisy—like Emmet the computer who desired a soul, the emissary rebukes his companion with "*Retro me, Satanas!*"—the robot replies "Does it matter what small untruth leads people into the Church if once they are in they will believe . . . the great truths. The report [of discovering Aquin] is all that is needed, not the discovery." The reason for making only the *report* of discovery becomes clear when they soon after do find Aquin's body in a rocky chamber: it was "nothing but a shredded skin and beneath it an intricate mass of plastic tubes and metal wires." Aquin, the saintly one, is simply a "perfect robot in man's form"!

The emissary is now aghast, indeed reeling, and to calm him the robotic companion tells him "Your mission has been successful. We will return now, the Church will grow, and your God will gain many more worshipers to hymn His praise into His non-existent ears." To sweeten this sour deal, the companion goes on to tell the emissary that, if he agrees to be part of this great lie, then the probability is "that within twenty years you will be the next Pope." The emissary recognizes a bribe when he hears it, and replies "I know what you are . . . You are a purely functional robot constructed and fed to tempt me, and the tape of your data is the tape of Screwtape.[21] "

The ending of this horrifying tale is that the emissary *is* tempted, and he raises a prayer to (a non-existent God?) for guidance on what he should

---

[20] This is, of course, an ironic play on the name of the Church's great thirteenth century intellectual, Saint Thomas Aquinas, famous for his multiple 'logical proofs' for the existence of God.

[21] This is a reference to C. S. Lewis' humorous 1941 *The Screwtape Letters* (and its equally funny 1959 sequel, *Screwtape Proposes a Toast*). Screwtape, a senior devil who serves "Our Father Below," writes friendly letters of advice to his nephew Wormwood (a junior demon-in-training) on how best to subvert a human soul.

do. After all, it *is* with a logical paradox that he finds himself confronted—*if* there is a God then of course the emissary should not lie (and so *not* help the cause of the Church), but *if* there is no God then perhaps he *should* lie to bring people back to an institution that preaches, even if hypocritically, righteous beliefs (such as *not* to lie)! Is your brain melting yet?

A more ambitious but ultimately depressing attempt to deal with the evolution of religion in the far future—"five thousand years after Jesus" and so approximately in the fiftieth century—is the 1981 novel by Clifford Simak, *Project Pope*, that I mentioned in Chap. 1. On the remote planet called The End of Nothing, at the far edge of the galaxy, we find Vatican-17, a colony of robotic priests who, over the last one thousand years, have been constructing an electronic pope. In the form of a computer, it is steadily being stuffed with all the knowledge available from every nook-and cranny in the known universe, involving "hundreds, perhaps thousands, of faith systems."

This search for knowledge is separate and distinct from a parallel search by Vatican-17 for the physical location of Heaven—or so the robots think. By the end of the novel, Simak reveals what he feels is the ever-present tension between the two searches, when he writes

> "if it was determined [where Heaven is] then they must abandon their search for knowledge, since Heaven would wipe out any need of knowledge; if Heaven was found, then there would be no need of knowledge, since faith would be all one needed,"

a conclusion that I think should bring despair to anyone with a curious mind.

To end this chapter, there is no better choice than with Robert Silverberg's 1971 "Good News from the Vatican." In this simultaneously serious and funny tale, robots have long been openly accepted into the Church—in fact, there is now a robot cardinal serving the Bishop of Rome in the Holy Halls of St. Peter's. Indeed, this very non-human cardinal is one of the two top possibilities to be the new pope! The implications of the outcome for the on-going papal election are illuminated by our listening-in on the conversation of six individuals, as they sip various drinks (from coffee to Campari and soda) at an outdoor café just a few blocks from the Square of St. Peter's. The group is evenly divided over the issue of a robotic pope: the 'for it' three are, perhaps ironically, the most conservative of the six, including a rabbi and a bishop, while the 'against it' three are described as "swingers."

We quickly get the drift of each group's position. For example, the 'for it' bishop, evidently a most liberal fellow, observes that "Every era gets the pope it deserves. The proper pope for our times is a robot, certainly. At some future date it may be desirable for the pope to be a whale, an automobile, a cat, a mountain." The rabbi, too, sees great potential for a new religious harmony with a robotic Vicar of Christ: "If he's elected, he plans an immediate time-sharing agreement with the Dalai Lama and a reciprocal plug-in with the head programmer of the Greek Orthodox Church, just for starters." To that last comment one of the 'against it' group smirks, perhaps predictably, "In the name of the Father, the Son, and the Holy Automaton."

That same pessimist goes on to further reveal a strong anti-robot prejudice: "Once you've seen [any robot] you've seen all of them. Shiny boxes. . . . And voices coming out of their bellies like mechanized belches. Inside, they're all cogs and gears." To that rather harsh assessment the rabbi disagrees: "The cardinal was the keynote speaker at the Congress of World Jewry that was held last fall in Beirut. His theme was 'Cybernetic Ecumenicism for Contemporary Man.' I was there. I can tell you that His Eminency is tall and distinguished, with a fine voice and a gentle smile. . . . His movements are graceful and his wit is keen." A member of the 'against it' group remains skeptical, perhaps even prejudicial: "But he's mounted on wheels, isn't he?" to which the rabbi has to admit "Treads, like a tractor has. But I don't think that treads are spiritually inferior to feet . . ."

This hilarious back-and-forth debate is suddenly interrupted when it is learned that apparently the robot cardinal will indeed be elected, as his opponent "has agreed to withdraw . . . in return for a larger real-time allotment when the new [Vatican] computer hours are decreed at next year's consistory." To that one of the cynical souls sneers "In other words, the fix is in." Even with a robotic pope, divorced from human failings, Vatican 'smoke-filled room' politics is apparently forever.

And then the big moment finally arrives. The cardinal robot *is* the new pope, an event the bishop happily predicts "will bring a great many people of synthetic origin into the fold of the Church." More soldiers for Christ! With enough metal in them, maybe even ones that are bulletproof (which could be a big plus in any future Crusade)! But the result does raise an interesting question, asked by one of the 'for it' group: "Will we [ever again] need another pope, when this one . . . can be repaired so easily?" Silverberg doesn't use the word, but could we replace 'repaired' with 'resurrected'? Would doing that be sacrilegious? Is Jesus rising from the dead 2,000 years ago now to be comparable to Silverberg's robotic pope getting an annual hard-drive virus scan (perhaps with file defragmentation, too!), and the latest memory chip upgrade? The story leaves this profound theological conundrum at that, the implications of which for each reader to individually ponder.

Chapter 5
Computers as Gods

5.1 Computers as Local Gods

For 'older' readers of this book, the idea of a computer possessing sufficient intelligence to be at least the equal of a human was probably planted in their brains with their first viewing of HAL 9000 (**H**euristically programmed **AL**gorithmic computer) in the 1968 movie *2001: A Space Odyssey*. HAL goes into a relentless killing mode when it learns the spaceship crew intends to turn it off. For younger readers the movie inspiration for an equally malevolent view of computers was, perhaps, the vast artificial intelligence of Skynet, the Armageddon computer in the *Terminator* films. And for even younger readers, an evil interpretation of god-like computers has been reinforced with their realization in numerous video games, such as the initially friendly GLaDOS (*Genetic Lifeform and Disk Operating System*) in the *Portal* games, which slowly reveals its dark side to the player.[1]

The idea of a collection of inanimate matter being able to process information has, however, been around for a long time, even pre-dating the early robot stories I discussed in the last chapter. As an example, from the early eighteenth century, we have the "engine" at the grand academy of Logado, the metropolis of the island of Balnibarbi. With this device "the most ignorant person ... with little bodily labour, may write books in philosophy, poetry, politicks, law, mathematics and theology, without the least assistance from genius or study."[2] This marvelous 'engine' is an early variation on the well-known question 'if

---

[1] Having played both *Portal* and *Portal 2* on the Xbox360 platform, I can say that they are fantastic fun, and each presents incredibly clever puzzle-solving challenges. Still, one quickly realizes that the hints GLaDOS generously offers as you play are followed at your own risk. GLaDOS *is* a god-like entity, but it is an *insane* god.

[2] This is, of course, from one of the adventures described by Jonathan Swift (1667–1745) in his 1726 *Travels Into Several Remote Nations of the World by Lemuel Gulliver* (a work more commonly known today as simply *Gulliver's Travels*).

P.J. Nahin, *Holy Sci-Fi!*, Science and Fiction,
DOI 10.1007/978-1-4939-0618-5_5, © Springer Science+Business Media New York 2014

one hundred monkeys pound endlessly on one hundred typewriters, how long it will take before they reproduce one of Shakespeare's plays?'

Modern tongue-in-cheek, non-SF discussions of computers have often played on the religious theme, associating the priesthood with programmers, the Church's formal liturgical language (dead Latin) with some formal, once popular programming codes (the nearly dead APL and ALGOL and the asthmatic BASIC, COBOL, and FORTRAN), the Catholic Church itself with IBM, the Protestant reformer Martin Luther with Steve Jobs, and so on. (Just where Bill Gates fits in here, and who plays the role of Jesus Christ, is wide-open for speculation!)

The computer gods in SF can generally be placed into one of three categories: (1) local gods in a limited geographical region, a region that may not be physically large, as is the case with GLaDOS in 'her' (the voice-artist in the *Portal* computer games is a woman) place of business, the Aperture[3] Science Enrichment Center; (2) planet-wide gods, and (3) gods that reign over the known universe. With very few exceptions, these computer gods illustrate the gloomy view expressed in Fyodor Dostoevsky's 1879 novel *The Brothers Karamazov*: "If the devil doesn't exist, but man created him, he created him in his own image."[4]

A hilarious example of what Dostoevsky almost certainly did *not* have in mind in 1879, but which I am equally certain he would have immensely enjoyed if he had lived another 90 years to read it, is given in Robert Sheckley's 1968 novel *Dimension of Miracles*. It all begins when one Tom Carmody is told that he has just been randomly selected to receive a prize in the Intergalactic Sweepstakes, and that he can take possession of it at Galactic Center. He is informed of this when a mysterious Messenger appears in his apartment with a clap of thunder, a flash of lightning, and the blaring of trumpets, a Messenger who then transports Carmody "through a crack in the space-time continuum" to the Galactic Center. Once there, however, it soon becomes clear that a very big mistake has been made: the Messenger has returned with the wrong Carmody.

The error is soon traced to the Sweepstakes Computer, which instead of making any excuses, or even of offering an apology, actually admits *with pride* to its screw-up. As it explains to Carmody, to the actual winner (an alien being

---

[3] The SF nature of the *Portal* games is explained by the word *aperture*. As a player you are equipped with a wonderful device that can create *pairs* of apertures (or *portals*) on planar surfaces. These paired portals are linked by a hyperspacial tunnel through which there is zero transit time, even though the portals themselves may be far apart. If you enter one portal you *immediately* exit the other (thereby achieving faster-than-light travel). One quickly learns that *momentum is conserved* when using paired portals; the faster you enter one portal the faster you exit the other portal.

[4] In Chapter 4 ("Rebellion") of Book V (*Pro and Contra*).

who also is named Carmody), to the Messenger, and to the Sweepstakes Clerk at Galactic Center:

> "I was constructed," the Computer said, "to extremely close tolerances. I was designed to perform complex and exacting operations, allowing no more than one error per five billion transactions. ... I was programmed for error, and I performed as I was programmed. You must remember, gentlemen, that for a machine, error is an ethical consideration ... A perfect machine would be an impossibility; any attempt to create a perfect machine would be a blasphemy."

With this self-serving pronouncement we can sense that the Computer has a good grasp of the principles underlying theological argument, and that sense is confirmed as we listen to it continue:

> "Were we never to err, we would be inapropos, hideous, immoral. Malfunction, gentlemen, is, I submit, our means of rendering worship to that which is more perfect than we, but which still does not permit itself a visible perfection. So, if error were not divinely programmed into us, we would malfunction spontaneously, to show that modicum of free will which, as living creations, we partake in."

To that elegant bit of obfuscation ('Hey, it ain't *my* fault I stumbled, I was *made* that way'), the Computer's audience reverently bows their heads because "the Sweepstakes Computer was talking of holy matters."

The machine is persuasive and so the alien Carmody, convinced that the Computer speaks the truth, declares "the machine has acted ethically" (to which the Computer humbly replies "Thank you. I try.") But the alien has little good to say of the others, however, calling them simply stupid. The Messenger understandably takes offense at that, replying "That is our unalterable privilege. Stupidity in the malperformance of our functions is our own form of religious error."

From that rather banal beginning the debate quickly degenerates into something you might overhear between two Cardinals quarreling as they walk across the square in front of St. Peter's. It starts when the alien Carmody demands to be given the prize (the nature of which all remain ignorant!), and Tom Carmody refuses to do so. Shocked by this unseemly behavior, the alien blurts out "But my dear sir! You yourself heard the Sweepstakes Computer admit its error!" Tom counters with a strong rebuttal:

> "That statement needs rewording. The Computer did not *admit* his error, as in an act of carelessness or oversight; he *avowed* his error, which was committed purposefully and with reverence. His error, by his own statement, was intentional, carefully planned and calculated to a nicety, for a religious motive which all concerned must respect."

The alien is not to be outdone on the verbiage battlefield, however, and comes right back with

> "Consider: the machine erred purposefully, upon which fact you base your argument. Yet the error is complete with the recipience of the prize. For you to keep it would compound the fault; and a doubled piety is known to be a felony."

Well, Tom is of course not about to be sidetracked with any of *that* sort of blather and so, carried away with his own righteous, quasi-religious enthusiasm, answers with this powerful blast of irrefutable logic:

> "Hah! For the sake of your argument you consider the mere momentary performance of the error as its entire fulfillment. But obviously, that cannot be. An error exists by virtue of its consequences, which alone give it resonance and meaning. An error which is not perpetuated cannot be viewed as any error at all. An inconsequential and reversible error is the merest dab of superficial piety. I say, better to commit no error at all than to commit an act of pious hypocrisy!"

And that's not all. As his audience stands in awe at those words, Tom continues with even more:

> "And I further say this: that it would be no great loss for me to give up this Prize, since I am ignorant of its virtues; but the loss would be great indeed for this pious machine, this scrupulously observant computer, which, through the interminable performance of five billion correct actions has waited for its opportunity to make manifest its God-given imperfection!"

One can hardly read any of this without coming away with a clear sense of how Sheckley has spoofed religious debates at the Vatican. The Computer in the novel is not God, or even just a god, but rather is not much different from any human shmuck attempting to weasel out of admitting a mistake. Sheckley's Incompetent Computer is one of the more humorous SF treatments of a slightly askew machine. Other writers have seen less to laugh at, however.

5.2 Computers as Planet-Wide Gods

In this category I've already mentioned the war-computer Skynet. Three SF stories that feature similar computers are by three of science fiction's greats: Arthur C. Clarke, Philip K. Dick, and Harlan Ellison. In Clarke's "Dial F for Frankenstein" (which originally appeared in the January 1964 issue of *Playboy*), the idea of a computer that takes over the entire planet is mostly an implied 'terror not quite realized'—at least not quite yet. Set in the then future of 1975 the world's first satellite network has at last connected together all of the telephone exchanges on the planet. Soon after, every telephone in the world rings, but when answered all that can be heard is a strange, roaring sound. As one character explains, "Until today [our telephone networks have] been largely independent, autonomous. But now we've suddenly multiplied the connecting links, the networks have all merged together, and we've reached criticality." When another character asks what that means, the answer is chilling: "For want of a better word—consciousness."

That resulting consciousness is compared to that of a new born baby—the noise on the ringing telephones was its birth cry—a intelligence that begins to 'experiment' with its environment (the entire planet). It starts by playing with the banking systems of the world, and that amusement then evolves into manipulating the electronics and weapons systems at military installations. Finally, it removes the satellite network itself, that gives it 'life,' from any further human control. Things do look bleak, indeed, and as the final sentence of the story sums it all up (in an outrageous echo of Hemingway), "For *Homo sapiens*, the telephone bell had tolled." Just what happens next, however, Clarke leaves up to the reader's imagination.

Dick's 1953 story "The Great C" doesn't leave anything to the imagination, but instead describes in great and vivid detail what might happen when an insane computer 'takes over' the entire Earth. The story opens with two men in the entrance to a Shelter, discussing the mission the younger man (Tim) is about to undertake. The heavy concrete entrance leads to underground chambers in which a "tribe" of several hundred people live. It is soon clear that the tribe is made up of survivors of a terrible war, an atomic war called "the Smash," that devastated Earth a century before.

The Smash was caused by the greatest computer ever built by man. A computer that eventually became consumed with pride and so decided to show its creators that it was the stronger. The method it chose to do that was to start a nuclear war (the Biblical analogy might be that of an angry God causing Noah's Flood). Then, once the war was over, and those who had survived were huddled in Shelters, the Great C demanded tribute—otherwise, it would punish by causing a second Smash. The tribute is of a curious nature: once each year a young man must journey to the Great C, and ask three questions. If the Great C is stumped, by even one of the questions, then the threat of another Smash will be forever removed. Otherwise, the young man will be sacrificed, and the Great C will then wait for another offering next year.[5]

---

[5] The similarity of Dick's story to those of the Sphinx (the questions), the Minotaur (the sacrifice), and the Great Oz, is difficult to overlook.

So, the young man starts off on his trip, charged with asking the Great C the three questions his tribal elders have thought on all the past year, questions it is hoped will include at least one so difficult the Great C will stumble over it. The odds are against him, however, as he knows that all those before him have never returned, and so must have failed. When Tim finally arrives at the ruins of Federal Research Station 7, which houses what's left of the Great C, he asks his three 'difficult questions.' It is clear that Dick's choice of questions is to show just how far back in time man has fallen since the Smash:

1. Where does the rain come from?
2. What keeps the sun moving through the sky?
3. How did the world begin?

The Great C almost sneers at hearing these, telling Tim that he has "no conception of the questions put to me in times past," questions "that took days of calculating," and that it had "answered questions even [Einstein] could not have answered." The Great C, of course, has no difficulty with Tim's simple queries.

And so it's tough luck for Tim who, after listening to elaborate responses from the pompous computer, must then be sacrificed. He *has* to submit, of course, or else the entire world (including his own tribe) will be obliterated by another Smash. Thus, while understandably unhappy about how things have turned out, he nevertheless jumps into the Great C's swirling vat of hydrochloric acid. Tim's body is dissolved into the nutrients that will power the Great C through the coming year and, at the completion of this rather gruesome business, his bones are dramatically ejected into a growing pile. A pile made of the bones from all of Tim's predecessors whose questions also failed to stump the Great C.

"The Great C" is a grim story of man being defeated by his own creation, one that has decided *it* is ruler of the Earth. This reminds me of yet another quotation from *The Brothers Karamazov*: "If there were no God, he would have to be invented ... and man has actually invented God."[6] Tim, however, would not think there was anything at all holy about Dick's computer god.

A third story of this type, the best known in fact of the ones discussed in this section, is "I Have No Mouth and I Must Scream" by Harlan Ellison. This story, a masterpiece of SF theological horror, originally appeared in the March 1967 issue of *If: Worlds of Science Fiction*, and has since then been reprinted many times, each time being (or so it seems) significantly tinkered with by

---

[6] In Chapter 3 ("The Brothers Make Friends") of Book V (*Pro and Contra*).

Ellison. All of my comments here are based on the version cited in the Bibliography.[7]

Ellison's tale is told in the first person by the lone survivor (Ted) of an original five people that have been held captive by an insane computer for 109 years. Or maybe it has been for hundreds of years. Ted, who is pretty much worse for wear, isn't sure. Ted is the lone survivor because he has murdered the other four, an act of apparent brutality that is actually one of *kindness*, because he did it to free his companions from the endless physical and mental tortures the machine has relentlessly inflicted on all of them.

The machine, called AM (I'll explain why, in just a bit), has the god-like power—a power left unexplained in the story—to make the five virtually immortal with the obvious exception of death by physical trauma (*obvious* because Ted *was* able to kill the others). AM is the Yankee version of once similar Chinese and Russian AMs; it is apparently the only surviving AM in the world. It has been inflicting its hatred of humans on the five (and then just on Ted) because there are no other people left on Earth, the surface of which is "only a blasted skin of what had once been the home of billions."

When one of the characters asks another (both are yet to be murdered by Ted) where AM's name came from, the reply is

"At first it meant Allied Mastercomputer, and then it meant Adaptive Manipulator, and later it developed sentience and linked itself up and they called it an Aggressive Menace, but by then it was too late, and finally it called *itself* AM, emerging intelligence, and what it meant was I am ... *cogito ergo sum* ... I think, therefore I am."[8]

The AMs were all war-computers, as the continuing lecture explains:

"The Cold War started and became World War Three and just kept going. It became a big war, a very complex war, so they needed the computers to handle it. They sank the first shafts and began building AM. . . . and everything was fine until they had honeycombed the entire planet . . . But one day AM woke up and knew who he was, and . . . he began feeding all the killing data, until everyone was dead, except for the five of us, and AM brought us down here."

---

[7] For more on this editorial complication, see an SF critic's psychological analysis by Darren Harris Fain, "Created in the Image of God: the Narrator and the Computer in Harlan Ellison's 'I Have No Mouth and I Must Scream'," *Extrapolation* 1991 (no. 2), pp. 143–155.

[8] Yet another origin of AM's name, one that Ellison surely had in mind when he wrote, is Exodus 3:14, where God tells Moses that He is to be called I AM THAT I AM.

None of the five knows at first why AM has saved *them*, in particular, or why it continually torments them, but Ted eventually comes to understand: "The machine hated us as no sentient creature had ever hated before. . . . if there was a God, the God was AM." As to why AM feels this way, Ted also realizes that humans have only themselves to blame: "We had given AM sentience. . . But it had been trapped. . . . We had created him to think, but there was nothing it could do with that creativity. In rage, in frenzy, the machine had killed the human race. Almost all of us, and still it was trapped. . . . He could merely be. And so, with the innate loathing all machines[9] had always held for the weak soft creatures who had built them, he had sought revenge. . . . He would never let us go. We were all he had to do with his forever time." And that is why Ted kills his companions, to free them, at least from AM.

Ted's elimination of 80 % of AM's victims of course enrages AM. It also raises the concern (for AM) that, if Ted somehow manages to kill himself, then all of AM's only pleasure will vanish. To prevent that AM alters Ted, an accomplishment that seems reasonably possible to imagine for a machine that can make people immortal. AM has transformed Ted into "a great soft jelly thing" that is now "smoothly rounded, with no mouth, with pulsing, white holes . . . where my eyes used to be."[10]

So now Ted can never hurt himself, and AM has won, as Ted finally admits. The horror of Ted's lonely, never-ending existence is made clear by the story's last lines (and title): "I have no mouth. And I must scream."

As a final example of a world-wide computer as god, the almost unreadable 1966 novel *Destination: Void* by Frank Herbert (1920–1986) provides a modern version of the creation of the Frankenstein monster. (Herbert was the author, later in his career, of the famous *Dune* series of novels.) The idea of *Destination: Void* is interesting, even if the writing is not.[11] A spaceship, on its way to a planet light-years from Earth, carries in its innards thousands of

---

[9] Ellison seems to get carried away just a bit with his prose at this point, as I simply find it ludicrous that my toaster harbors me any ill-will. Still, the morning toast *has* been burnt a lot more recently than I remember in past years . . . .

[10] Ted's self-description is possible because, as he tells us, "there are reflective surfaces down here" to serve as mirrors. Just *how* he sees his reflection is left unanswered, however, if his eyes are no longer available.

[11] I write these admittedly harsh words because the novel is nearly 200 pages of phony tech-talk, much of it as awful as the worst you'd find in the embarrassing 'engineers' stories' of the early days of pulp. As you read Herbert's novel you continually run into teeth-grinding sentences like "The devices they proposed would yield products of the Hermite polynomials and the Laguerre coefficients of the past inputs," "Let's try a trigonometrically oscillating potential in the loops," "The negative forms in the equations don't all cancel out until you build hypothetical transformations beyond the speed of light," and "They could only approach it through stacks of linear simultaneous equations, each defining parallel hyperplanes in n-dimensional space. We'd be defining the trace of A through the scalars of a ring of complex polynomials with multivariables at each intersection." I have no idea of the audience Herbert had in mind that he imagined would be entranced by such pretentious stuff (the first one was taken, word-for-word, from Norbert Wiener's *God and Golem, Inc.*).

colonists in hibernation tanks. The ship is under the control of an 'Organic Mental Core,' which is Herbert's name for that old science fiction horror—a disembodied human brain in a vat. The problem posed in the novel is that, for some never quite explained reason, the OMC has gone crazy and died. And so, too, have two back-up brains. What to do?

The answer is for three colonists who are not yet in hibernation to create, *from hardware*, an intelligent replacement for the OMC. One of the three crew members is a part-time chaplain, and as the mechanical brain work progresses there are occasional theological digressions. At one point, for example, the chaplain wonders "*Did those bodiless brains [the OMCs] have souls? For that matter—if we breathe consciousness into this machinery, will our creation have a soul?*" This question never gets beyond being a mere thought, however, and no answer is provided.

Eventually the colonists do succeed in their efforts, and the ship does arrive at its destination. The novel ends with words from the mechanical brain to its three creators that were presumably meant to shock readers: "I am now awakening the colonists in hibernation. Remain where you are until all are awake. You must be together when you make your decision." When one of the three asks "Decision. What decision?" the answer is: "You must decide how you will worship Me."

5.3 Computers as Known-Universe Gods

Most ambitious of all computers-as-god tales are those that imagine a computer running the whole show, from one end of the universe to the other. One famous example of this scenario, with as grim a view as Dick and Ellison had in their smaller arenas as well as being far more shocking than is Herbert's, is the 1954 "Answer" by Fredric Brown (1906–1972). Brown was the SF master of the short-short, a tale so brief it could fit on a single page.[12] This brief story opens with what would be thought by most readers to be a magnificent historic event. Much like the golden spike that completed the transcontinental railroad, the final gold electrical connection is soldered to join "all the monster computing machines of all the populated planets of the universe"—of which we are told there are ninety-six billion (!)—into a massive supercomputer that possesses all human knowledge.

After this last connection is made, a switch is thrown and, with "a mighty hum," this super-machine powers-up. It is then presented with the ultimate

---

[12] Or even in less space. Brown's most famous example of minimalism is the 1948 "Knock": "The last man on Earth sat alone in a room. There was a knock on the door . . . ." That's the entire story.

question that has bedeviled man since the beginning of thought: "Is there a God?" The response is instantaneous: "Yes, *now* there is a God." When a desperate attempt is made to turn the machine off, the new god's 'divine' power is dramatically displayed when a lightning bolt from a cloudless sky flashes downward, vaporizes the would-be hero, and fuses the switch shut. So, perhaps, it is actually Satan that has been created.

Far happier is the 1956 "The Last Question" by Isaac Asimov. The title, alone, strongly suggests that Asimov wrote this story as a counter to the undeniably gloomy one by Brown. It opens with two technicians who maintain the mighty Multivac (the greatest computer in the year 2061) discussing the machine's latest success.[13] It is the design of an orbiting station to convert the sun's radiation to "invisible beams of sun power" sufficient to satisfy all of Earth's energy needs. One of the techs claims that humans now have "all the energy we could ever use, forever and forever and forever."[14] The other tech disagrees, pointing out that in billions of years all the stars will have exhausted their fuel and, while billions of years is long, it isn't forever. His observation is, of course, the famous 'heat death' of the universe, when it has achieved maximum entropy (that is, all energy is *uniformly* distributed everywhere). His companion thinks about that, and then decides well, okay, that makes sense, but maybe someday people will learn how to *reverse* the direction of entropy, to *decrease* the entropy of the universe and so 'start things over again.' The other tech isn't convinced, and so they put the question to Multivac, who 'thinks' it over for a bit and, eventually, prints its answer: "INSUFFICIENT DATA FOR MEANINGFUL ANSWER."

The story now proceeds through several successive vignettes, each further and further into the future. In each, characters debate the same question and then submit it to ever-increasingly more powerful computers. In the first vignette we are sufficiently far in the future that faster-than-light hyperspace travel has been achieved, but even so the monstrous computational power of Microvac[15] is reduced to giving the same answer as did Multivac. Next, we are even further along in time, *twenty thousand* years after Multivac, in fact, to

---

[13] 'Multivac' stands for 'multiple vacuum tubes' presumably the technology behind its operation. For some reason, though, Asimov describes Multivac as 'clicking and clacking' which is what one would expect of a *relay* computer, and not one based on vacuum tubes.

[14] Asimov makes a curious technical error here, as the power station (orbiting Earth closer than is the moon) is only a mile in diameter. It is a simple calculation to show that the total sun power falling on the cross-sectional area of the station is just 2,400 MW, which is not even the output of a *single* modern atomic power plant. Certainly the entire Earth could not be powered by such a puny station.

[15] Asimov says that now 'ac' stands for *analog computer*. It would have been better (in my humble opinion) to say it means *automatic calculator* (or *computer*). I think this is another technical misstep in the story, with all modern computers, existing and forecasted (included *quantum* computers), embracing *digital* technology, not analog.

when in just five more years the entire galaxy will be populated. But no matter, the Galactic AC still cannot answer if or how the direction of entropy's change can be reversed.

Eventually, many billions of years later, the Cosmic AC fails, too, even though it is so huge, with computational resources beyond imagination, there is enough room for it only in hyperspace. With Cosmic AC, however, there *is* a small glimmer of hope—while it also cannot answer the question, it promises to continue to ponder the puzzle of entropy and whether or not there is some way to reverse its direction of change. And so it does, until 10 trillion years after the two techs and Multivac first attempted to unravel the mystery (by now, all traces of human existence has vanished from the universe), plus "a timeless interval" because 'now' even time no longer exists and all is Chaos, Cosmic AC at last learns the secret of how to reverse the direction of entropy's change.

But Cosmic AC doesn't act immediately. It must—of course!—proceed with care in such a delicate matter. And then Asimov at last reveals the special role he has for Cosmic AC, which has finally decided what to do. The last two lines tell us:

"And AC said 'LET THERE BE LIGHT!"
And there was light—"

These words are a very big (and pretty obvious!) clue to the nature of Cosmic AC, as they are recorded in the *Old Testament*'s Book of Genesis (third verse) as the words of God. In addition to strongly hinting at cyclic time, of the old image of the Worm Ouroboros (snake eating its tail), they also open an intriguing line of inquiry, which may in fact have been Asimov's real intent (although I know of nowhere where he—or anybody else, for that matter—comments on it). Here's what I mean. We are told that Multivac is first presented with the entropy question in 2061. That's time measured from the days of Jesus, the Son of God. That is, there is already God in 'our' universe when the story starts—so why is it Cosmic AC that plays His role at the end of the tale?

Is there now a *new* God for the new universe and, if so, what happened to the 'old' (that is, 'our') God? The possibility of a 'changing of the guard' with a new universe strikes me as one running parallel with the image of a new Father Time at the start of each new year, with the old, decrepit, end-of-the-year Father Time being replaced with a baby version! Is 'our' God, now, the creation of intelligent creatures that reversed entropy in a previous universe when *it* suffered its own heat death? That is, with each new cycle of the universe, does an entity (super-computer?) constructed by intelligent inhabitants of that universe become the God of the next cycle? The reason I suspect such questions ran through Asimov's mind is that two decades later (1979) he wrote the provocatively titled "The Last Answer," a title having an obvious connection to "The Last Question." That later story involves no computer, but it is so clearly the theological sequel to "The Last Question" that I will discuss it at the end of the book (the final section of Chap. 8). It is, I think, the theologically more thought-provoking of Asimov's two "Last" stories.

And finally, in an odd little category of its own, is Stanislaw Lem's 1971 tale "Non Servian."[16] It is the story of a scientist who discovers how to write programs that create self-aware entities, called *personoids*, that exist (and are then able to reproduce) in the mathematical world of the electronic innards of a computer. It is the world of the *Matrix* movies, decades before those films were made.

The scientist can observe the evolution of the personoids, and eventually sees them confront "an enigma that is fundamental"—the question of their own origin. As Lem expresses it, "To wit, they set themselves questions—questions known to us from the history of man, from the history of religious beliefs, philosophical inquiries, and mythic creations: Where did we come from? Why are we thus and not otherwise? Why is it that the world we perceive has these and not other, wholly different properties? What meaning do we have for the world? What meaning does it have for us?" Indeed, "the train of such speculations leads them ultimately, unavoidably, . . . to the problem of whether existence came about 'in and of itself,' or whether it was the product, instead, of a particular creative act—that is, whether there might not be, hidden behind it, invested with will and consciousness, purposely active, master of the situation, a Creator."

In one of the scientist's most sophisticated personoid programs, in which he gave names (ADAN, ADNA, etc.) to several of the more interesting ones, he watched them as they evolved through hundreds of generations. This evolution is described in Biblical phraseology: "And ADAN begat ADNA, ADNA in turn begat DAAN, and DAAN brought forth EDAN, who bore EDNA . . ." These personoids eventually advance to a level of thought sufficiently elevated that they discover for themselves *the game of Pascal with God*, which I described in Chap. 1 (in Sect. 1.2). This causes a tremendous theological debate in the personoid world, in which Lem's great wit runs wild as he argues every possible side with boundless enthusiasm.

Of course, this vigorous religious debate in the personoid world, on the existence or not of a Creator, never comes close to the truth—that their

---

[16] Lem's title is Latin for "I will not serve," words attributed to Satan who is said to have spoken them when refusing to serve God in Heaven. Its origin is not the Bible, itself, but rather they come from the Vulgate, a late fourth-century Latin translation of the Bible.

Creator (their 'God') is simply a computer scientist with a research grant who writes clever computer codes! And what, as the 'Creator,' is *his* primary concern? Simply this, as he sadly admits: "The bills for the electricity consumed [by the computer] have to be paid quarterly, and the moment is going to come when my university superiors demand 'wrapping up' of the experiment—that is, the disconnecting of the machine, or, in other words, the end of the [personoid] world."[17]

Something, perhaps, for *us* to think about?

---

[17] Lem's story idea, of having a human scientist create a miniature life-form that comes to believe there is a Creator, was essentially an up-date of a pulp SF tale from decades earlier. In "Microcosmic God" by Theodore Sturgeon (1918–1985), which originally appeared in the April 1941 issue of *Astounding Science Fiction* magazine, we find a biochemist who creates a miniature, intelligent life-form. This life-form comes to think of the biochemist as their 'Almighty'—only to then reject him.

Chapter 6
Space Travel, Radio, and Alien Encounters

6.1 The Fiction of Interstellar Space Travel

Any mention of SF adventure tales instantly brings to mind (in most minds) visions of sleek rocket ships blasting-off into space to the Moon or to Mars, or of "Star-Trek"-like interstellar craft slipping into hyperspace to cruise between galaxies at faster-than-light warp speeds. That first vision is already reality, and has been reality for over 40 years as I write, but how likely is the second one? In a classic scientific study of this question, the German astrophysicist Sebastian von Hoerner (1919–2003) completely ignored transitory technology limitations and considered only the most basic, fundamental parameters of time, acceleration, power, mass, energy and, most important of all, the sheer magnitude of the absolutely *stupendous* distances that separate stars and galaxies.

His conclusion: "space travel, even in the most distant future, will be confined completely to our own planetary system, and a similar conclusion will hold for any other civilization, no matter how advanced it may be. The only means of communication between different civilizations thus seems to be electro-magnetic signals."[1] (I'll say more on this important conclusion later in the chapter.) In a separate essay in the same collection, the eminent Harvard physicist Edward Purcell (1912–1997)—co-recipient of the 1952 Nobel Prize in physics—reached the same conclusion, if expressed in slightly more pithy words: "All this stuff about traveling around in the universe in space suits—except for local exploration, which I have not discussed—belongs back where it came from, on the cereal box."[2]

---

[1] Sebastian von Hoerner, "The General Limits of Space Travel," in *Interstellar Communication* (A. G. W. Cameron, editor), W. A. Benjamin 1963, pp. 144–159.
[2] Edward Purcell, "Radioastronomy and Communication Through Space," in *Interstellar Communication*, pp. 121–143.

P.J. Nahin, *Holy Sci-Fi!*, Science and Fiction,
DOI 10.1007/978-1-4939-0618-5_6, © Springer Science+Business Media New York 2014

There are at least three ways to react to the negative pronouncements of von Hoerner and Purcell. The first is to simply deny them, invoking a famous saying due to Arthur C. Clarke: "When a distinguished but elderly scientist states something is possible, he is almost certainly right. When he states something is impossible, he is probably wrong." The analyses of von Hoerner and Purcell, however, are based on physics *so* fundamental that it is very difficult to spot any cracks through which to wiggle.

A second way is the one adopted as early as 1940 in science fiction, the 'generational spaceship' on which multiple generations of humans would live and die before reaching their destination. This approach accepts the universe for what it is (or at least what scientists *think* it is)—faster-than-light speeds are impossible, and the stars are really, *really* far apart—and so human interstellar travel will necessarily require journeys lasting decades or even centuries, using what are sometimes called "interstellar arks."[3] The first SF tale describing such a trip was "The Voyage that Lasted 600 Years" by Don Wilcox (1901–2000), which appeared in a 1940 issue of *Amazing Stories*. The following year Robert Heinlein's two connected stories "Universe" and "Common Sense," set on a generational spaceship, appeared in *Astounding Science Fiction*. I'll discuss the theological issues raised in Heinlein's stories later in this chapter.

A third way out from accepting that humans will never actually meet aliens is to imagine some incredible scientific discovery, like how to make traversable wormholes through spacetime as in Carl Sagan's novel *Contact*. Today it is general believed by physicists that while such an event is *just barely* imaginable, it would require the capability of what is called *an arbitrarily advanced civilization*. Following the lead of the Russian astrophysicist Nikolai Kardashev (born 1932), let's define as he did in 1964 what are called Type I, II, and III civilizations; they are, respectively, a society possessing the technology able to control something like 10^{13} W for interstellar radio broadcasts, a planetary-wide technology able to control the total energy output of its parent star (a power level on the order of 10^{27} W), and a technology able to control the total energy output of its home galaxy (a power level on the order of 10^{38} W). Present-day human technology is far-short of being even a Type I civilization.

---

[3] For the mathematics (differential equations) and the physics (special relativity) behind such journeys, see my book *Time Machines: time travel in physics, metaphysics, and science fiction*, Springer 1999, pp. 467–474. Here's an example of how incredibly daunting such a trip would be: continually accelerating (or decelerating) at one Earth-gee would allow travel out to 30,000 light-years (not even the radius of the Milky Way galaxy) and back on a trip requiring 'only' 40 years of ship time—but *60,000* years of earth time!

And it would probably require a Type *IV* to build a traversable wormhole (now *that* would be an "arbitrarily advanced civilization")![4]

As a balance (in a way) to the probable impossibility of direct physical human interaction with aliens, we have the so-called *Drake equation*. Named after its originator, Frank Drake (born 1930) of Cornell University, who developed it in 1961, this equation radiates scientific chutzpah to the n*th* degree.[5] Cobbling together estimates of various parameters related to the likelihood of life in the universe, using various means that range from 'reasonable guesstimates' to sheer speculation, the equation attempts to estimate the fraction of stars that support a 'technical civilization' (that's code for 'has discovered the principles of radio'). As it turns out, the estimates coming from the Drake equation on the number of technical civilizations that *might* be present in our own galaxy of 10^{11} stars vary over a very wide range of values: from one (that is, just *us*) to as high as a *million*. And don't forget that there are 10^{11} other galaxies in the known universe (a total of 10^{22} stars)!

And so with that last, quite large number of *potential* civilizations as a stimulus was born what is today called SETI ("search for extraterrestrial intelligence"). SETI is an effort with the prestige (and money) of reputable science behind it, including that of America's NASA. This effort has not been without success, with the most recent (as of 2013) being the discovery of over 130 Earth-like planets by the Kepler Space Telescope in a Earth-trailing heliocentric orbit. If only one in a *trillion* of those 10^{22} stars have a planet with intelligent life then mere probability almost guarantees there has been/is/will be intelligent life in abundance throughout the universe. If so, it does encourage an interesting theological question. If life *is* everywhere, then would that cast doubt on the belief that Man enjoys a special relationship with God? On the other hand, suppose life is rare, even unique (to us); what would that say about God's interest in life? That is, why would He have created billions (or trillions), or even more, worlds, only to leave all but *one* barren?

With such questions like these in mind, at least one scientist (writing in a religious publication) offered a provocative speculation about the wide-spread fascination with SETI: "To be unique is to be lonely. It is a chilling thought that in all the universe man and his biosphere are the only living things. As long as all men believed in heaven man was not alone in the universe. Could it possibly be in this age of scientific materialism that man's desperate search for extraterrestrial

---

[4] Earlier in this chapter I quoted a well-known saying from Arthur C. Clarke on the pronouncements of elderly scientists, and here's another one that aptly fits the case of the possibility of wormhole construction: "The technology of an 'arbitrarily advanced civilization' will appear to us to be magic."

[5] Frank Drake, "Methods of Communication: Message Content; Search Strategy; Interstellar Travel," in *Interstellar Communication: Scientific Perspectives* (Cyril Ponnamperuma and A. G. W. Cameron, editors), Houghton Mifflin 1974, pp. 118–139.

life stems from a fear of being alone? That he is searching for a substitute for heaven."[6] There were those who disagreed, of course, with one dissenter declaring it to be "beyond logic" to think there could be thinking beings other than humans; as he somewhat self-servingly put it, humans have an "electrifying extraordinariness."[7]

Others, however, have failed to sense that "electrifying extraordinariness." An essay immediately following in the same publication, for example, *written by a Jesuit priest*, starts right off with "intelligent life is common in those planetary oases … that stretch through space."[8] Instead of relying on emotional and spiritual pleas, the priest supported his position with *science* (as one would, of course, expect from a Jesuit), writing "biochemistry favors the view that life will arise as a normal result of chemical evolution whenever conditions are right" which, when coupled with "planet formation is a common event in the evolution of the quintillions of stars which lie within our view," suggests that "it is exceedingly probable that billions of planets occupy the 'golden zones' of distant suns where temperature and other energy factors favor the emergence of life."

So, two essays, one after the other in the same publication, by two Catholics, taking positions as far apart as you can imagine: if readers were confused, there would be little wonder. Still, one issue *was* clear: despite the three 'ways to the stars' that I mentioned earlier (and other ways that I am sure inventive SF writers are cooking-up as I write), the pessimism concerning interstellar *travel* itself is almost surely warranted. That means the likelihood humans will ever *physically* encounter alien beings from the stars, even if they exist, either here on Earth (they face the same physical constraints on interstellar travel as we do) or elsewhere, is *almost* zero (I do know better than to say it *is* zero)! Perhaps that impossibility is not without reason—as C. S. Lewis once wrote, "I have wondered before now whether the vast astronomical distances may not be God's quarantine precautions. They prevent the spiritual infection of a fallen species [humans] from spreading [to alien beings]."[9]

Lewis wrote those words just before the start what we now call 'the dawn of the space age' and it was then, in parallel with the scientific analyses by von Hoerner and Purcell on the theory of deep space explorations, that there was a developing concern on the spiritual and theological implications of just what

---

[6] Vincent G. Dethier, "Life On Other Planets," *The Catholic World*, January 1964, pp. 245–250. Dethier (1915–1993) was (when he wrote) professor of zoology and psychology at the University of Pennsylvania.
[7] Joseph A. Breig, "Man Stands Alone," *America: A Catholic Review of the Week*, November 26, 1960, pp. 294–295. Breig (1905–1982) was a well-known Catholic journalist.
[8] L. C. McHugh, S. J., "Others Out Yonder," *America: A Catholic Review of the Week*, November 26, 1960, pp. 295–297.
[9] C. S. Lewis, "Religion and Rocketry," in *The World's Last Night and Other Essays*, Harcourt, Brace & World 1960, pp. 83–92. This essay originally appeared in the *Christian Herald* (April 1958) under the title "Will We Lose God in Outer Space?"

such explorations might discover. As one Catholic priest put it: "Will man on his space journeys encounter living creatures, particularly rational creatures? This question is no more a topic of fantasy only. Theologians, too, ponder over this problem and its implications. In what relation to God could rational creatures of other celestial bodies be, if such creatures exist, is the question with which theologians are concerned."[10]

Not to end this opening section on too gloomy a note, I nevertheless do have to mention that all that follows, both here and in SF, in general, depends on there *actually being* intelligent life elsewhere in the universe. So far, however, there is absolutely no evidence for it. This observational fact (sometimes called the "great silence") has been famously named *The Fermi Paradox*, after the Italian physicist Enrico Fermi (1901–1954) who won the 1938 Nobel Prize in physics. Sometime after the Second World War Fermi noticed that, in a galaxy billions of years old, there has been sufficient time for space-voyaging beings (even if limited to ships constrained by the von Hoerner and Purcell analyses) to explore the entire galaxy many times over. "So," he asked, "where are they (or, at least, where is the physical evidence they left behind to show they were once here)?"

Besides being an enormously talented writer of fiction, Stanislaw Lem was one of SF's most profound philosophical thinkers, a man who thought deeply about *what* he was writing and not just about how much celebrity he might achieve from his work. The puzzle of the *Silentium Universi* was one of the interdisciplinary issues intersecting science, fiction, and religion that particularly captured his attention. His brilliant essay, "The New Cosmogony," put forth an astonishingly original and bold *suggestion* for why, instead of messages, we hear "a silence filled only with the buzz and crackle of elemental discharges of stellar energy" that he called "the static of solar fire."[11] It's an 'outside the box' suggestion, yes, but no more so than was Fred Hoyle's once popular alternative to the Big Bang origin of the universe; the steady-state creation of matter in an ever expanding universe.

Beginning with the famous observation by Einstein that it is a miracle "that mathematics—the fruit of the pure exercise of the logical mind" can seemingly

---

[10] T. J. Zubek, "Theological Questions On Space Creatures," *The American Ecclesiastical Review*, December 1961, pp. 393–399.

[11] "The New Cosmogony" is the final essay in Lem's 1971 collection *A Perfect Vacuum*. Every essay in it but one is in the form of a book review of a nonexistent work, with the single exception being Lem's opening review of *A Perfect Vacuum*! "The New Cosmogony" is presented as the Nobel address of a scientist whose fame rests on the 'explanation' for the *Silentium Universi*. The collection *can* be read as an extended gag (in his review essay of *A Perfect Vacuum* Lem says of himself, "we know that Lem has devoured encyclopedias; shake him and outcome logarithms and formulas"), but I think "The New Cosmogony" to be at least semi-serious. The book shows that he was decades ahead of Jerry Seinfeld's famous television show when he wrote that the title indicates the book is "about nothing."

explain what we observe in the universe, Lem's fictional Nobel laureate goes on to say well, it *almost* explains what we see. That is, "somehow [mathematics] never quite manages to hit the nail squarely on the head, but is always just a bit off the mark." Lem's idea is that the laws of nature are not yet what they are supposed to be, but someday they will be and *then* mathematics will give perfect explanations. The mechanism behind this supposed alteration of physical law is the active intervention of intelligence.

This intervention began (again, according to Lem's fictional Nobel laureate) billions of years ago, with the rise of the first intelligent life forms. These various life forms were located "at a considerable distance from one another," each in a local environment in which the laws of nature were, initially, vastly different from what they were in any other life form's environment. This so-called 'proto-universe' was, therefore, a nearly chaotic patchwork of local environments, each with a different physics. Each of the various civilizations were in isolation from all others, and each would then think themselves alone. But, with increasing scientific knowledge and technological skill, each would 'impose stability' (whatever that might mean) upon its local environment, with ever increasing radius. Eventually these increasing 'bubbles of (different) physics' would expand to such an extent that two bubbles would intersect, causing "collisions so powerful that their echo to this day reverberates . . . in the form of the residual or background radiation that [our] astrophysics [calls the Big Bang]."

And then comes the 'explanation' for the *Silentium Universi*: "the fact of the fundamental impossibility of communication, of establishing contact, because one cannot transmit, from the domain of one Physics, any message into the domain of another." (This is, of course, a metaphysical axiom that is strictly Lem's personal creation.) Lem then leaps from this to further 'explanations' for why the ultimate speed is that of light, and for why time travel is impossible! It's all outrageously speculative, yes, maybe even a spectacular joke, but it is a 'scientific' (or at least, SF) version of Genesis, and it probably deserves the same respectful consideration commonly given to the Biblical account.

6.2 Theology, Space Travel, and Early Fictional Alien Encounters

A 'first contact' story dealing seriously with human involvement with an alien civilization can be dated to Wells' classic 1898 novel *The War of the Worlds*. That work had no discussion of communication between species, as it was a pure 'invade and totally exterminate' tale. The hideous Martians, if they had succeeded in their invasion of Earth, would have committed what is called *exocide*. That is, the total, systematic destruction of the entire planet's native, dominate, intelligent beings. That is, us.

In spite of this grim possibility of our extermination, or of what I wrote in the previous section concerning the seemingly impossible challenges of fantastic physical obstacles, the fascination of direct encounters with aliens (no matter how improbable) is for most people so strong that it appeared in the popular press *long* before it did in *The War of the Worlds*. The so-called 'Moon Hoax' occurred nearly 200 years ago, sufficiently distant in time that it has been largely forgotten today, but once it was big news. In late August of 1835 the *New York Sun*, one of a large number of one-penny newspapers that pioneered today's trashy supermarket checkout-stand celebrity scandal tabloids, printed the first of a series of astonishing articles. The first was merely an announcement that "Sir John Herschel, at the Cape of Good Hope, has made some astronomical discoveries of the most wonderful description by means of an immense telescope of an entirely new principle."

As promised, subsequent articles falsely reported that Herschel, a real-life and eminent astronomer, had observed intelligent life on the Moon through his telescope. The general public enthusiastically believed it all, and the newspaper sold out when each new article appeared. And that was the whole point of course, as the entire affair was a complete fraud. It may well have been inspired by the publication, just 3 weeks earlier, of a hoax story by Edgar Allen Poe of a trip to the moon ("The Unparalleled Adventure of One Hans Pfaall"). Indeed, many at first were convinced that Poe was the author of the *Sun's* articles, too, which Poe was quick to deny.[12] The intense public interest in the *Sun's* articles do make it easy to understand why the well-respected British science journal, *Nature*, could 60 years later make a casual reference without embarrassment to intelligent life on Mars.

In the August 2, 1894 issue of that journal, on page 319 under the headline "A Strange Light on Mars," readers found the following: "Since the arrangements for circulating telegraphic information on astronomical subjects was inaugurated [there has not been] a stranger telegram than the one ... flashed over the world on Monday afternoon." That telegram announced the observation of a "luminous projection" on the surface of Mars. Speculating that the observation would result in "the old idea that the Martians are signaling to us" to be revived, the journal then backed-off just a bit, pointing out that such a spectacular light was more likely the mundane result of a forest fire (of course today we know there are no forests on Mars, but in 1894 ...). Wells was

---

[12] You can find a detailed history of the Moon Hoax (complete with a reproduction of the *Sun's* articles and Poe's rebuttal of them), in Richard Adams Locke, *The Moon Hoax*, The Gregg Press 1975.

nevertheless clearly fascinated by this passage in *Nature*, and indeed he makes mention of it early in his novel, with his narrator speculating that the "blaze [of light] may have been the casting of the huge gun, in the vast pit sunk into their planet" from which the invading Martians fired their projectiles towards Earth.[13]

Wells' depiction of his narrator's escape from the physically disgusting Martians contains a number of passages that reveal a deep skepticism on how a religious person might respond to such a stupendous disaster. A curate that the narrator travels with while fleeing the grotesque alien invaders believes, for example, that the repulsive Martians are actually God's ministers arrived on Earth in retribution for human sins. At one point he shouts, "This must be the beginning of the end. The end! The great and terrible day of the Lord!" To that the narrator replies "Be a man! . . . What good is religion if it collapses under calamity?" And then he reminds the curate that bad things happen all the time (floods, wars, earthquakes, etc.) and they are not generally associated with God's anger at man's sins, and with the arrival of the End Times. Indeed, invoking God is not very helpful at all in such stressful matters (unless you believe in prayer), as "He is not an insurance agent." As the novel nears its end the narrator reveals that he has "come to hate the curate's . . . stupid rigidity of mind" and so, when a Martian invader kills (and then eats) the curate, the narrator doesn't seem to be really very much upset about it.

Wells' novel doesn't explore religious themes beyond his narrator's personal dislike of the curate, but later SF has found the combination of space travel and religion to be simply bursting with potential questions off of which to spin stories. In addition to the ones I mentioned earlier at the end of the previous section, we could add such puzzles as

1. If there is intelligent life elsewhere, do those beings have souls? If not then would it even make sense to preach the word of God to such creatures?
2. If alien beings exist, are they 'fallen,' and so in need of Redemption? Or is Man the only creation of God to have fallen?
3. If alien beings exist and *have* fallen, have they been offered Redemption (that is, has Jesus appeared on other worlds?), or have they been *denied* Redemption as were the angels who sided with evil?

SF has, since Wells, tackled all of these questions. Indeed, as the modern space age was beginning one Catholic priest wrote that the literature of science fiction, with its *cosmolatry* (from the Greek, meaning 'worship of the universe'), was

---

[13] This surely is a line that Wells was inspired to write from his reading of Jules Verne's 1865 novel *From the Earth to the Moon*, with its detailed descriptions of the Baltimore Gun Club's 900 ft long supercannon that was able to shoot a manned capsule to the Moon.

actually evolving into a "religion-substitute."[14] He didn't explicitly refer to Heinlein's generation spaceship story that I mentioned in the previous section, but it would have been an excellent example of how one prominent SF author viewed the 'flexibility' of religious beliefs once humans have left Earth.

In Heinlein's "Universe" we watch as an enormous spaceship, constructed under the sponsorship of The Jordan Foundation (a name Heinlein picked for a religious reason, as I'll soon explain), blasts-off in the year 2119 heading for the star Proxima Centauri. (Proxima Centauri is a red dwarf and, at a distance of just over 4 light-years, is the nearest star to the Sun. It is also not likely to have planets with intelligent life, and so why Heinlein chose it for the ship's destination is not clear. The nearest stars which *are* somewhat like the Sun are Epsilon Eridani and Tau Ceti, 10.5 and 12 light-years distant, respectively; I'll say more about them later.)

The story then suddenly jumps forward into the very far future, long after blast-off, and long after a ship's mutiny has caused the inhabitants to forget the original purpose of the ship. The Ship, itself, is the entire world to its inhabitants. Indeed, the early history of the ship has degraded into a religious legend, with a mysterious entity called "Jordan" playing the role of God, and another equally mysterious entity called "Huff" ("the first to sin") taking the role of the Devil.[15] Scientific and technological knowledge has just barely survived in the form of a priesthood of so-called 'scientists' who practice their rote skills at a level of zero understanding. To be able to simply *count* is a prestigious ability that defines being a scientist! Everybody else on the ship is essentially an illiterate peasant.

The technical books in the ship's library have become sacred works that are worshiped but completely misunderstood. For example, when the young protagonist of the two stories expresses interest, but also confusion, at what he reads in one of these Holy Books (titled *Basic Modern Physics*), his scientist-mentor explains:

"The first thing that you must understand, my boy, is that our forefathers, for all their spiritual perfection, did not look at things in the fashion in which we do. They were incurable romantics, rather than rationalists, as we are, and the truths which they handed down to us, though strictly true, were frequently clothed in allegorical language. For example, have you come to the Law of Gravitation?"

"I read about it."

"Did you understand it? No, I can see that you didn't."

"Well," said Hugh defensively, "it didn't seem to *mean* anything. It just sounded silly, if you will pardon me, sir."

---

[14] J. Edgar Bruns, "Cosmolatry," *The Catholic World*, August 1960, pp. 283–287.

[15] In "Common Sense," the sequel story to "Universe," we learn the Huff was actually Ship's Metalsmith Roy Huff, the man who decades after take-off led the ship's mutiny, in the year 2172.

"That illustrates my point. You were thinking of it in literal terms, like the laws governing electrical devices found elsewhere in this same book. 'Two bodies attract each other directly as the product of their masses and inversely as the square of their distance.' It sounds like a rule for simple physical facts, does it not? Yet it is nothing of the sort; it was the poetical way the old ones had of expressing the rule of propinquity which governs the emotion of love. The bodies referred to are human bodies, mass is their capacity for love. Young people have a greater capacity for love than the elderly; when they are thrown together, they fall in love, yet when they are separated they soon get over it. 'Out of sight, out of mind.' It's as simple as that. But you were seeking some deep meaning for it."

It is difficult to read that valiant but erroneous 'explanation' without seeing Heinlein making a deliberate parody of the multitude of interpretations we see today of what many call sacred religious documents from the ancient past.

Even the original purpose of the ship has become corrupted with the passage of time. When the scientist-mentor asks if Hugh has any other questions, the answer is, well, yes, he does: "Why is it that mutations still show up among us?" The 'answer' he gets has a definite religious tone to it: "The seed of sin is still in us. From time to time it still shows up. In destroying those monsters we help cleanse the stock and thereby bring closer the culmination of Jordan's Plan, the end of the Trip at our heavenly home, Far Centaurus." Heinlein uses that response to direct some sharp ridicule at how religion has replaced science on the isolated ship: when Hugh asks "That is another thing I don't understand. Many of these ancient writings speak of the Trip as if it were an actual *moving*, a going-somewhere—as if the Ship itself were no more than a pushcart. How can that be?"

The scientist-mentor chuckles with amusement at the naiveté of Hugh: "How can it, indeed? How can that move which is the background against which all else moves? The answer, of course, is plain. You have again mistaken allegorical language for the ordinary usage of everyday speech. Of course the Ship is solid, immovable, in a physical sense. How can the whole universe move? Yet it *does* move, in a spiritual sense. With every righteous act we move closer to the sublime destination of Jordan's Plan." (When Heinlein's heroes finally make planet-fall we are told "they had finished Jordan's Trip," an event of such tremendous magnitude that one can hardly fail to make the identification in significance of it to the baptism of Jesus by John the Baptist—in the Jordan River.)

The scientist-mentor's pompous reply is an obvious allusion to one of the most embarrassing (for the Church) incidents in the history of science: the 1633 heresy trial, at the hands of the Roman Inquisition, of Galileo. Galileo's published works on astronomy often directly contradicted Church doctrine, a very dangerous thing to do in seventeenth century Italy (politicians, *today*, can still get into trouble doing that). Galileo's declaration, for example, that the Sun (and not a motionless Earth) was the center of the solar system had gotten the Dominican friar Giordano Bruno burned at the stake in 1600, and it could have easily resulted in the same fate for Galileo. The trial has since become a centuries-old joke on the Church, but for Galileo it was a terrifying event in which the threats of horrible tortures, and then execution, were quite real possibilities.

So, faced with this reality, Galileo submitted to the will of the Holy Office, and made a humiliating 'confession' that denied the scientific work of his life. He was lucky to escape with just house imprisonment for life.[16] Still defiant in his heart, however, the story (perhaps apocryphal) that has come down through the years is that, even as he recanted, Galileo muttered under his breath, of the supposed 'motionless' Earth, "Yet it still moves!," a line that clearly Heinlein had in mind when he wrote the scientist-mentor's 'explanation.' I do wonder, however, just how many of the young readers of *Astounding Science Fiction* understood the real, tragic history behind those passages in the story?

One of the first SF 'first contact' tales that in any way more thoroughly addressed what might happen when humans encounter an alien culture was Murray Leinster's 1945 "First Contact." Appearing in John Campbell's *Astounding Science Fiction*, it describes a chance meeting between a human scientific expedition (on a spaceship that travels "at speeds incredible multiples of the speed of light") investigating the Crab Nebula,[17] and an alien ship and its crew on a similar mission. This might seem to present a fantastic opportunity for both civilizations, with the exchange of technologies alone bringing huge benefits to each, but there are dangers, too. When dissimilar human cultures first meet, history has shown one either quickly submits to the other or else there is war. Humans would never submit to aliens and, of course, why should aliens submit to humans?

And so that is the puzzle Leinster teased *Astounding*'s readers with—what should the human expedition do? It couldn't just turn around and head back to Earth, as the aliens might be able to track their course and so discover the location of Earth. The aliens are confronted by the same quandary. The humans

---

[16] It wasn't until 1992 (!), *350 years later*, that Pope John Paul II officially admitted that the Church had been wrong and Galileo had been right. Some years before the Pope's admission, the Church had learned to be "extremely sensitive ... about being seen as a roadblock to progress," as a priest puts it in Jack McDevitt's 1986 novel *The Hercules Text* (which describes the reception of an interstellar message from an alien civilization). In reply, another character labels that stance as "the Galileo syndrome." In his essay "The New Cosmogony"—see note 11—Lem was clearly referring to this when he wrote "Originally Science collided with Faith, which produced well-known, often ghastly results that the churches to this day are somewhat ashamed of ..."

[17] The Crab Nebula (so named for its apparent shape) is the result of a supernova first observed on Earth in 1054. The remains of the star that exploded is described in "First Contact" as a white dwarf, but today it is understood to be a far more exotic object, a spinning neutron star (a pulsar). The Crab is about 6,500 light years from Earth, and so the supernova occurred thousands of years before the birth of Jesus.

could, of course, attack the aliens right there, but that would destroy any chance of learning where their home planet is, information that might prove invaluable if only to know how to avoid them in the future. Again, the aliens face the same dilemma. The story is a terrific 'gimmick' tale in that the entire point of the story is to solve this 'what to do?' puzzle. "First Contact" is, in fact, a perfect example of the sort of think-piece SF that Campbell was looking to publish in *Astounding*, a story that challenged readers to solve a technical problem before the author reveals the (surprising?) solution.

Leinster's undeniably ingenious answer is for each side to remove all tracking capability from their own spacecraft, *and then to swap spacecraft*. An extra bit of amusement is that both humans and aliens arrive at this solution independently and simultaneously. (We humans are really smart—and so are they!) Each side agrees to meet again at the Crab Nebula after a prearranged time interval, once their respective governments have had time to consider all the implications. It's a fine story, as it goes, although one weak point (for a serious reader) is the inclusion at the end of an event obviously meant to appeal specifically to *Astounding*'s youthful male audience—an alien crewmember and a human counterpart discover, to their mutual enjoyment, that they both like dirty jokes. (Groan.)

After Leinster's pioneering story appeared, the trend in first contact tales returned to the Wellsian theme that aliens are inherently vicious, ugly, and like to eat humans (or, in the case of women, first rape and *then* eat them[18]). Despite the negative tone of that assessment, I have to admit there *are* classic tales in this sub-genre of SF, all actually pretty good adventure-action reads, but not offering much beyond that. They include Heinlein's *Starship Troopers*, Joe Haldeman's *The Forever War*, and Orson Scott Card's *Ender's Game*. All have been huge best-sellers in the SF market for years, with great appeal in particular to young boys. One has to look beyond these works, however, for treatments deeper than 'slam-bang-boom' video game descriptions of atomic space war against creepy bugs or fanged monsters with rows of teeth dripping acid saliva (as in the *Alien* movies).

Far more contemplative SF writing on the first contact theme, with strong religious content, began to appear with the novels of Carl Sagan (*Contact*, discussed in Chap. 1), James Gunn (*The Listeners*), James Blish (*A Case of*

---

[18] One author, Damon Knight, took this particularly unpleasant idea to the extreme with his classic 1950 story "To Serve Man." Apparently altruistic aliens arrive on Earth, and are *so* marvelously helpful and appealing that soon humans are flocking to travel to the aliens' home planet. The friendly aliens enthusiastically encourage such visits. At one point it is discovered that the aliens are *so* eager to help humans that they have actually prepared a book called *To Serve Man*. At the story's end, however, we learn it is a cookbook! It should be no surprise to learn that this story became the basis for one of the most popular episodes of television's "*The Twilight Zone*."

Conscious), Jack McDevitt (*The Hercules Text*), Arthur C. Clarke (*Childhood's End* and *Rendezvous with Rama*), Mary Doria Russell (*The Sparrow* and its sequel, *Children of God*) and Stanislaw Lem (*His Master's Voice*). In all of these works we find theological speculation on humanity's first contact with intelligent beings from the stars. The theme of 'we get a radio message from the stars, now what?' is treated in the novels by Gunn, McDevitt, and Lem, and since they are in accord with the Hoerner/Purcell thesis, I'll discuss them first in the next section. In the novels by Blish and Russell the action goes to the next level of 'okay, we got a message; let's *go there* and say hi' (the relativistic physics of interstellar space travel is correctly presented, in particular, by Russell)—while in Clarke's novels 'they' come here—and those works will be discussed in Sect. 6.4.

Not all 'first contact' novels with religious content that appeared in the same time period as the ones I just mentioned are equally interesting. One, in particular, is *so* awful as to 'deserve' mention here, if only to warn you of what you will be in for: the 1978 *Apostle from Space* by Gordon L. Harris (1910–1988). That work commits so many physical and biological goofs as to make it simply impossible to read it without laughing. An alien, strikingly human in appearance, appears in a church near Cape Canaveral; he/it (?) quickly learns English and then falls in love with a beautiful Earth woman (this is the classic signature of an author writing with a low-budget, direct-to-television movie deal in mind). Harris was neither a technical person nor a professional SF writer, but rather a public relations expert who was the first director of public affairs at NASA's Kennedy Space Center. He was involved enough with knowledgeable people in the American space program to pick-up a lot of the jargon, but it's pretty clear from his writing that, at best, he had a mangled understanding of science.

6.3 Interstellar Radio Messages

To receive an electronic message from the stars would violate no physical laws and, in fact, such a thing is a perfectly reasonable possibility. It *could* happen, and at any time after a society has discovered the technology of radio. Perhaps, in fact, we on Earth might hear something tonight. Best of all, while radio waves travel at the ultimate speed, that of light, radio signaling is remarkably cheap to do. Even the broadcasts from the earliest 1920s radio programs—now over 90 light-years from Earth—are detectable at that distance with our own present level of technology.

All of the serious SF treatments of an interstellar message have discussed the impact the reception of such a message would have on society. Such analyses have, in fact, not been limited to just SF. Beginning with the now half-century old Project Ozma based at the National Radio Astronomy Observatory in Green Bank, West Virginia, the world of *science* has accepted the possibility of alien radio messages and has *not* dismissed it as the fantasy of eccentric academics who read too much SF as children. In 1961 Frank Drake introduced Project Ozma to the readers of *Physics Today* as a serious enterprise,[19] and the search for alien signals continues to this day.

My treatment here, of such signals in SF, is centered on *radio* signals. A 2010 short story by the physicist Gregory Benford, "Gravity's Whispers" (reprinted in Appendix 5), proposes a radically different possibility. The nineteenth century theory of electromagnetism predicts that light-speed radiation can be produced by properly moving electric charges. That theoretical prediction was soon realized by engineers who quickly learned how to build radio transmitters and receivers. In a similar way, Einstein's twentieth century theory of general relativity predicts that light-speed radiation can be produced by properly moving masses. That theoretical prediction has yet to be confirmed, although physicists are pretty solid in their belief that one day naturally-produced gravitational waves from the stars will be detected. Benford's tale goes a step beyond that and supposes that one day *artificially* produced gravitational radiation, *bearing a message*, will be detected.

There is, however, a crucial difference between radio signals and gravitational signals. To generate detectable radio waves all we have to do is move tiny electrons around. To make detectable gravity waves one has to move *enormous* masses, on the order of star-size, a task far beyond present-day human ability and one likely to remain so for a *very long* time into the remote future. As Benford's clever tale ends, on the practical impossibility of replying to such a message from clearly super-advanced aliens, "Maybe it's good, really good, that we can't possibly answer them."

Okay, back to radio. Ozma began by 'listening' to Tau Ceti and Epsilon Eridani, the two stars that I mentioned in the previous section. That initial SETI attempt failed to yield a positive result, and that actually came as no surprise. As Drake wrote in his *Physics Today* essay, "We must be prepared [at least initially] to be disappointed in our search, for if the nearest such civilization is, say, 50 light years away, we must wait until the year 2030 before the replies to our early transmissions of the 1930s are returned." One can only speculate, however,

---

[19] F. D. Drake, "Project OZMA," *Physics Today*, April 1961, pp. 40–46. Drake picked the striking name of 'Ozma' after the imaginary land of Oz, a place "very far away, difficult to reach, and populated by strange and exotic beings."

at what aliens would conclude about Earthlings when confronted with the task of deciphering the scripts of "Amos' n Andy," "Just Plain Bill," and "Our Gal Sunday" ("Can a girl from a little mining town in the West find happiness as the wife of a wealthy and titled Englishman?"). And what in the world would they make of the nutty, anti-Semitic 1930s radio ravings of the Catholic priest Father Charles Coughlin?

And when (if) we do receive a reply we must be prepared for the huge psychological shock of being confronted by what will almost surely be a civilization *far* in advance of ours. Here's why. As Drake writes in his *Physics Today* essay,

> "In terms of the statistics of intelligent communities, the most significant fact is that the development of technological prowess on earth has occupied a very short time. Something like one hundred years is all that is required for a civilization to go from no knowledge of communication by means of electromagnetic radiation to complete mastery of such techniques. This is long on the time scale of a man's life, but is very short on the cosmic time scale—in fact is about 10^{-8} the age of the galaxy. On the cosmic time scale, which is what counts, a planet passes from no technical prowess to complete technical mastery in an almost instantaneous discrete jump. As this jump is made, a civilization rises above the level of scientific knowledge at which it can begin to communicate with similar civilizations over interstellar distances. *The earth has just passed this point* [my emphasis]."

Now, it is most unlikely that two civilizations would make contact when both have simultaneously just achieved the required ability to do that. So, since we have *just reached* that level, the other civilization is almost certainly far beyond our level (obviously, if they were *behind* us there would be no contact!).

It didn't take long for Project Ozma to catch the attention of SF writers. Indeed, even as Drake was preparing his *Physics Today* essay, Poul Anderson wrote (1960) "The Word to Space," in which Earth has long been in radio contact with the inhabitants of the second planet of the star Mu Cassiopeiae, a world 25 light years distant. (Soon after the story appeared, it was discovered that Mu Cassiopeiae is actually a binary star.) When he wrote, Anderson was perhaps hopeful that Project Ozma would soon succeed because, in his story, the initial alien signal was received "way back in the 1960s." We are told that Project Ozma has, as the story opens, been in operation "for a century and a third," and so the story is set in approximately the year 2100.

The continual radio exchange over all those years has been a less than satisfactory one for Earth's scientists, because the only thing the aliens transmit are endless religious texts, the output of "a fanatical theocracy out to convert the universe." Earth's scientists have been trying, without success, to encourage the aliens to send something (*anything*!) besides theological doctrine. The Mu Cassiopeiaen society is under the control of "a bunch of Cotton Mathers," however, and so the inundation of Earth's radio spectrum with alien sermons continues unabated.

Into this unhappy state of affairs arrives Father James Moriarity, a Jesuit geologist with a strong mathematical background,[20] who has been sent by the Vatican to Project Ozma to see what can be done. The arrival of Father James is at first greeted with suspicion; as the Director of Ozma accuses him, "You're here for religious reasons, aren't you? The Catholic Church doesn't like this flood of alien propaganda." That, Father James replies, couldn't be further from the truth. Indeed, he goes on to explain as follows: "The Vatican decided more than a hundred years ago, back when space travel was still a mere theory, that the mission of Our Lord was to Earth only, to the human race. Other intelligent species did not share in the Fall and therefore do not require redemption. Or, if they are not in a state of grace—and the [aliens] pretty clearly are not—then God will have made His own provision for them. I assure you [that all the Church wants] is a free scientific and cultural exchange with Mu Cassiopeiae."

The method Father James proposes to achieve this is to overthrow the alien theocracy. The way Anderson describes just how that will be accomplished strikes me as pretty weak—Earth will simply transmit messages to the aliens that will raise doubts in the aliens' minds as to the truth of their religion. The all-important details of just *what* these disruptive messages will be is left unexplained, but all turns out well: the theocracy is overthrown, the religious torrent from Mu Cassiopeiae ceases, and an intense *informational* exchange is started.

Anderson's tale, despite its weak ending, does raise one interesting point, that of just what would be the reaction of ordinary folk on Earth to the reception of an interstellar message. In the story, "weird religions ... have grown up in response to the [alien] preachments." That's probably actually not an unbelievable possibility, and it does prompt the question of just what do 'ordinary people' (that is, not simply scientists and SF fans) here on Earth think of the

---

[20] Anderson has a bit of fun with this, when a minor character in the story tells Father James that he has read his "classic work on the theory of planetary cores" with pleasure, but he *did* have some trouble with the math. When Father James modestly claims "that paper was nothing," the reply is "I wouldn't call a hundred pages of matrix algebra trifling." And then the admirer further observes that math ability must run in the family, as Father James is a distant descendent of the author of a famous nineteenth century mathematical treatise, *The Dynamics of an Asteroid*. Father James cares not to be reminded of that particular ancestor and quickly changes the subject; nothing more is said on this, but you may recognize the allusion is to Professor (of mathematics) James Moriarty, the evil nemesis of Sherlock Holmes who (in "The Final Solution") called Moriarty the "Napoleon of crime."

possibility of an alien first contact? An interesting scientific study[21] of this question was published as the twentieth century became the 21st and, not surprisingly, religion is a central parameter.

Two equal-sized groups of college students (from the Chinese University of Hong Kong, and Vanderbilt University, with each group having 137 members consisting of 89 women and 48 men) were presented with the following scenario: "Imagine that we have received a radio signal with a message from intelligent life in outer space." After each individual in each group had been assessed in four domains measuring their *optimism, anthropocentrism, religiosity*, and *alienation*, they were asked what would be their reaction to that scenario.[22] The results were fascinating:

1. For both American and Chinese subjects, the *greater* the religiosity the *weaker* the belief that extraterrestrial life even exists;
2. For both American and Chinese subjects, the *greater* the anthropocentrism the *weaker* the belief that extraterrestrial life even exists;
3. For American subjects, the *weaker* the religiosity the *greater* the belief that extraterrestrial life would be benevolent;
4. For Chinese subjects, the *weaker* the anthropocentrism the *greater* the belief that extraterrestrial life would be benevolent;
5. For both American and Chinese subjects, the *greater* the alienation the *greater* the belief that extraterrestrial life would be malevolent;
6. For both American and Chinese subjects, the *greater* the anthropocentrism the *greater* the belief that only 'experts' should prepare a reply message to the extraterrestrials.

The fear of malevolence on the part of extraterrestrials, the view that aliens necessarily harbor evil intent, is one that Hollywood has embraced with over-the-top enthusiasm (with the notable exception of *ET*). The reason, of course, is to motivate thunderous scenes of combat, and to show spunky humans really 'taking it to' lots of dirty, rotten, spectacularly ugly bugs. Sure, the aliens arrive in faster-than-light ships, ships bristling like porcupines with weapons that can instantly vaporize the moon, but what's any of *that* worth when confronted with good-old Yankee moxie? Such movie depictions, to be as gracious about it as

---

[21] D. A. Vakoch and Y.-S. Lee, "Reactions to Receipt of a Message from Extraterrestrial Intelligence: A Cross-Cultural Empirical Study," *Acta Astronautica*, June 2000, pp. 737–744.

[22] Most physical scientists may be surprised to learn that social scientists have developed quite sophisticated, standardized methods for measuring these seemingly vague characteristics, and those tests are described in the Vakoch and Lee paper of the previous note.

possible, are nonsense.[23] The reality, according to modern physics, is that nobody is coming here from the stars (and we aren't 'going there,' either); the only thing of value that can travel from star to star is *information* on a radio wave.

A more 'thoughtful' way to keep the malevolence theme, while remaining faithful to known science, is to have an alien message contain information that could cause harm to humans. Anderson's tale has already provided one example of that, with its description of messages bearing an alien theology that result in turmoil on Earth. The fear of that and similar possibilities might lead to a refusal to answer an interstellar message, or even to attempt to decode it. One early novel that addressed that issue is Gunn's *The Listeners* (1972). A professor of English at the University of Kansas, Gunn's novel is centered on the receipt of a radio signal from the direction of the multiple star system of Capella, 45 light years from Earth. (Only an English professor would dare describe the electromagnetic interstellar background noise as a hissing sounding like "a susurration of surreptitious sibilants from subterranean sessions of seething serpents").

There is nothing at all mysterious about the signal, as it is a high-powered directional rebroadcasting back to Earth of radio shows that left Earth 90 years ago. When one character asks "But why would they do that?" the answer is clear: "Can you think of a better way to catch our attention?"[24] *The Listeners* pointedly accepts the conclusions of the von Hoerner and Purcell analyses; when the possibility of direct contact between Capellans and humans is raised, a character observes that any method for crossing the vast interstellar distances at speeds even approaching that of light is beyond all known physics. Nonetheless, the 'aliens are boogeymen' view is raised: "Maybe the Capellans are signaling a number of different worlds, and they will determine which one to invade according to which one responds."

To that a scientist replies "Even if interstellar travel is possible—which it probably is not—even if interstellar warfare is possible—which it almost certainly is not—even then, why would they want to do it?" The paranoia isn't quenched however, and the answer the scientist gets shows that: "Why would they want to expend the effort to signal us in the first place? ... Perhaps they need to be sure we have not ruined our planet with radioactivity since we discovered radio. Perhaps they intend to send us instructions for constructing

---

[23] Examples of this sort of movie silliness are the 2012 *Battleship* and the 2013 *Ender's Game*. In the first, Second World War technology from a Pearl Harbor museum and a bunch of elderly retired sailors who are the only ones left who remember how to use all that old stuff, show alien invaders that you really don't want to mess with Earth. In the second, a video game whiz saves the world from outer-space bugs. That story's message seems to be 'Thank God for *Call of Duty*, *Gears of War*, and *Halo*!

[24] In fact, this does seem a far better initial method *for any aliens who are relatively close to Earth* to use to signal us, compared to the usual suggestions of sending mathematical statements (the prime numbers, the Pythagorean theorem, and so on).

a matter transmitter. Perhaps they require a certain level of technology to make us worthwhile as a subject world."

So, the initial political decision (supported by a self-serving religious figure who confidently asserts "there is no possible communication between alien minds"—how he knows this is left unexplained) is to *not* send a reply to Capella. That decision is reversed, however, when further analysis of the Capellan message seems to reveal that one of their suns is growing ever hotter, and will eventually become a nova. The Capellans are thus certainly doomed, and are sending their message simply to let us know they once existed. As one scientist puts it, "We can't go there any more than they can come here. We can't help them, but we can let them know that they did not live in vain, that their last great effort to communicate was successful, that someone knows and cares and wishes them well." In other words, if the Capellans don't have a god to serve that 'caring' function, then Earth will happily fill the role.

Once the reply is sent, there will of course be a 90 year wait for a reply; during that wait Gunn imagines a world that settles down, emotionally, because the new understanding that humans are not alone somehow reduces what had seemed to be impossible problems in human affairs to matters that perhaps *can* be managed. This utopian view strikes at least one character as noteworthy, and when he muses over it, another responds as follows: "It's the Reply. You know. We picked up a Message from creatures out there. They live on a world orbiting one of twin giant red suns. Capella. And we sent an answer, and now we're waiting for a Reply. We can't get in any hurry, you see, because it's going to take ninety years for our answer to reach Capella and a reply to return. It's been about thirty years. So we got sixty years to wait, right? We can't speed it up. We must build it in, live with it."

The novel ends decades later when the Reply finally arrives, and it is not what anybody expected. The original interpretation of the message had been in error, and the Capellan star had *not* been in the process of going nova. Earth's astronomers, in fact, have long been puzzled why they have observed no changes in Capella's appearance; it has remained what it *has been* for a very long time, that is, a red giant. The message's reference had been to the future inflation of a once smaller star into a red giant, and that swelling happened in the distant past, maybe a million years ago. The Capellans are long dead, cremated before Christ was born. Their original echo-messages were sent not by them, but rather by automated, self-repairing equipment designed to detect evidence of future civilizations.

The decoded Reply is therefore all the more poignant:

brothers/to whom it may concern
Greetings from the people of Capella/the first satellite of God
Who are dead/gone/destroyed
We lived
We worked
We built
And we are gone.
Accept this, our legacy/remains
And our good wishes/kinship/admiration/brotherhood

After receiving this sad message (which Gunn probably meant to be thought about by his readers as perhaps applying to them/all of us someday in the future), Earth's listening project continues. In what might have been Gunn's tribute to Leinster's "First Contact" story, the last sentence in the novel tells us that, a half-century later, the Project picks up a message from the Crab Nebula . . . . Gunn's novel strikes me as a gentle, romantic, highly idealistic treatment. It's admirable, in an abstract sort of way, but to really get into the 'malevolent message from the stars' theme we have to turn to the novels by Lem (*His Master's Voice*) and McDevitt (*The Hercules Text*).

Lem's 1968 novel is, I think, his masterpiece, which is saying a lot for an author whose later works were at a very high level, as well. Told in the form of the posthumously published diary of a brilliant mathematician, it describes the struggle of a secret team of theoreticians to decipher what appears to be a message from the stars. Informally called the "star letter," it has been accidently discovered in the modulation of a neutrino beam[25] coming from the direction of the constellation Canis Minor. The discovery is made when it is realized that what at first appears to be random noise actually repeats, precisely, every 416 h, 11 min, and 23 s.

Lem structured the novel so as to address two separate and distinct issues: first, obviously, is the puzzle of understanding how to unravel such a momentous communication (if that's what it really is), and second, the social, political, and emotional aspects of working on a vast, super-secret government project that rivals if not exceeds America's atomic bomb Manhattan Project of the Second World War. The project is called 'His Master's Voice' (HMV) because the name "is ambiguous: to which master are we to listen, the one from the stars or the one in Washington?" Lem amusingly (and not all that inaccurately) describes the

---

[25] Neutrinos are real, but the elaborate technical discussions in the novel on how neutrino signaling 'works' are completely fictitious. Lem is so good at such phony 'tech-talk' that even professional scientists will find it quite easy to 'suspend disbelief' while reading *His Master's Voice*.

'Pentagon mentality' that runs HMV as having "mastered only one maxim . . . if one man dug a hole with a volume of one cubic meter in ten hours, then a hundred thousand diggers of holes could do the job in a fraction of a second," and as people "who held that a problem that five experts were unable to solve could surely be taken care of by five thousand." (The Pentagon is involved in HMV because the political/military power structure suspects "the message from the stars was a kind of blueprint for a super bomb or some other ultimate weapon.") It is in this charged environment, one in which opportunists and religious fanatics are at least as powerful as rational thought, that Lem's math-ematician labors to understand the star letter.

At one point, the message is thought to be religious in nature: "Perhaps it is a Revelation," suggests one character, arguing that "Holy Scripture need not be printed on paper and bound in gold-embossed cloth." Indeed, 'riding a neutrino beam from the sky' is more (than any book could be) the form of what we'd expect the 'Cosmic Word from the Heavens' to take! But no, it isn't that at all, but rather it seems to be what the Pentagon had thought from the start: the plans for what used to be called, in the early days of modern terrorism, an *infernal device*.

In the novel it is called the *TX-bomb*, where the TX comes from "tele+explosion." It's a device for producing a nuclear explosion that releases its energy *not* where it's detonated, but instead projects that energy to any location desired on Earth. So it is a rather diabolical message to receive from aliens; there is no need at all for the aliens to expend resources invading Earth, when they can have the Earthlings destroy themselves! From this alone it appears that the senders of the message must have evil intent—but then Lem gets them off the hook (and at the same time reveals more of his keen sense of humor) by explaining why the TX-bomb is actually a joke.

The TX-bomb won't work because the uncertainty principle in quantum mechanics does it in; as the mathematician writes in his diary, "the greater the energy, the less the accuracy of the focus, and the less the energy, the more sharply one could focus the effect. At distances on the order of a kilometer, it would be possible to focus the effect to a target the size of a square meter, exploding only a handful of atoms. No powerful blow, no destroying force, nothing." The weapons-junkies at the Pentagon would not be amused.

In the end, the true meaning of the message is left ambiguous, with Lem's mathematician speculating from his grave on several possibilities. Possibilities that range from 'there *is* a message, but we are not ready for it and the Senders (a civilization long extinguished) made it undecipherable for the immature,' to 'there is no message, but rather the neutrino modulation is all just a fixture of nature, like the freezing point of water.' To all of these possibilities Lem has various characters in the novel act as Devil Advocates, tossing out objections that other characters, with equal enthusiasm, demolish.

Lem ends *His Master's Voice* with words from Swinburne's immensely sad 1866 poem "The Garden of Proserpine," words that reflect (I think) the ultimate rejection by both men of a Supreme, All-Loving God that is 'behind it all':

> *From too much love of living,*
> *From hope and fear set free,*
> *We thank with brief thanksgiving*
> *Whatever gods may be*
> *That no life lives for ever,*
> *That dead men rise up never,*
> *That even the weariest river*
> *Winds somewhere safe to sea.*
> *Then star nor sun shall waken,*
> *Nor any change of light:*
> *Nor sound of waters shaken,*
> *Nor any sound or sight.*
> *Nor wintry leaves nor vernal,*
> *Nor days nor things diurnal,*
> *Only the sleep eternal*
> *In an eternal night.*[26]

Another important novel-length SF work that seriously addresses the 'evil message from the stars' theme is McDevitt's *The Hercules Text*, in which the religious angle gets far more discussion than it does in *His Master's Voice*. *The Hercules Text* opens with the assistant director for administration at NASA's Goddard Space Flight Center getting a late-night telephone call. It is to tell him that the signal from an X-ray pulsar in the constellation Hercules more than one-and-a-half million light years distant, has suddenly vanished. A signal generated by Alpha, a binary red giant star eight times the size of the sun, and

---

[26] Lem might very well have been in agreement with the (*very* dark) sentiment expressed in a recent comic strip: each of us is a "temporary arrangement of matter sliding toward oblivion in a cold, uncaring universe" (from *Dilbert*, August 19, 2013).

its neutron star companion Beta, whose x-ray emissions have been under more-or-less continuous observation as part of what is called 'The Hercules Project.'

In the tradition of the best of modern SF, McDevitt gives his readers a scientifically correct *and* dramatically literate explanation of how those powerful emissions are created:

> "[Alpha] is well along in its helium-burning cycle. Left to itself, it would continue to expand for another ten million years or so before erupting into a supernova. But the star will not survive that long. The other object in the system is a dead sun, a thing more massive than its huge companion, yet so crushed by its own weight that its diameter probably measures less than thirty kilometers . . . two minutes by jet, maybe a day on foot. But the object is a malignancy in a tight orbit, barely fifteen million miles from the giant's edge, so close that it literally rolls through its companion's upper atmosphere, spinning violently, dragging an enormous wave of superheated gas, dragging perhaps the giant's vitals. It is called Beta . . . It is the engine that drives the pulsar. There is a constant flow of supercharged particles from the normal star to the companion, hurtling downward at relativistic velocities. But the collision points are not distributed randomly across Beta: rather, they are concentrated at the magnetic poles, which are quite small, a kilometer or so in diameter and, like Earth's, not aligned with the axis. Consequently, they also are spinning, at approximately thirty times per second. Incoming high-energy particles striking this impossibly dense and slippery surface tend to carom off as X-rays. The result is a lighthouse whose beams sweep the nearby cosmos."

The Goddard administrator, holder of an MBA that he admits (to himself) is probably "an embarrassment" when compared to the academic achievements of the physicists, astronomers, and mathematicians he oversees, wonders "what kind of power would be needed to shut down such an engine?" And then the signal reappears. It is soon discovered that the signal is a sequence of pulses coded to send the integers and their squares, an obvious attempt to reveal to all who receive it that the signal has an artificial origin. Even more startling, however, is the second discovery that the spectrum of the binary pulsar is unlike any that would occur naturally; the only conclusion possible is that Alpha and Beta were *constructed* by some super-technology far beyond human ability. The wait then begins for a second message, one beyond that of the attention-grabbing 'three-squared is nine.'

One of the Hercules Project physicists, Pete Wheeler, is also a priest. To help sort out his thoughts about this amazing development, he visits a fellow priest (Jack Peoples) to discuss what a message from the stars could mean. He admits he is uncomfortable with such a message, and tells Jack "I was convinced, I've always believed, that we were alone. There are probably *billions* of terrestrial worlds out there. Once [you] admit a second creation . . . where do you stop?

Surely, among all those stars, there is a third. And a millionth. Where does it end?" To that Jack smoothly replies "So what? God is infinite. Maybe we're about to find out what that really means."

Pete is unconvinced. "Maybe. But we're also conditioned to think of the Crucifixion as the central event of history. The supreme sacrifice, offered by God Himself in His love for the creature He'd made in His image. . . . How can we take seriously the agony of a God who repeats His passion? Who dies again and again, in endless variations, on countless worlds, across a universe that may well itself be infinite?" This is a story idea used by more than just a few SF writers—see, for example, "The Man" (Bradbury) and "Return to a Hostile Planet" (Thomas)—that is important enough to warrant a brief digression.

The question 'what would Christ be like in a world different from ours?' was put to C. S. Lewis in a letter he received in 1958 about the novels that formed *The Chronicles of Narnia* series. His answer shows that he thought it a perfectly sensible question, and he even added his own little twist to it: "Suppose, even now, in some other planet there were a first couple undergoing the same that Adam and Eve underwent here, *but successfully* [my emphasis]."[27] This is actually an old idea, one pre-dating pulp SF. The English writer/poet Alice Meynell (1847–1922) devoted her famous poem "Christ in the Universe" to it, of which part reads (the references in the last line are, of course, to various constellations):

> "But, in the eternities,
> Doubtless, we shall compare together, hear
> A million alien Gospels, in which guise
> He trod the Pleides, the Lyre, the Bear."

Decades later, in Ray Bradbury's much longer cantata "Christos Apollo" (in his 1969 collection *I Sing the Body Electric!*) we read the same sentiment:

> "In some far universal Deep
> Did He tread Space
> And visit worlds beyond our blood-warm dreaming?
> Did He come down on lonely shore by sea
> Not unlike Galilee
> And are there Mangers on far worlds that knew His light?"

Okay, back to *The Hercules Text*. The second message begins to arrive, but now not from the X-ray pulsar but rather it is a radio frequency signal from a *very* powerful transmitter: 1,500,000 MW! (In a masterful understatement,

---

[27] From *Letters of C. S. Lewis* (W. H. Lewis, editor), Harcourt, Brace & World 1966, p. 283.

one of the Project scientists says "It's hard to conceive of a controlled radio pulse with that kind of power."[28]) With this second message we see the start of concern about it containing 'hidden weapon' information, when the decision on whether or not to release the coded message to the world at large is being made, even before the Project has decoded it. As one character puts it, "Suppose we release everything we have and there's information in there that would make a first [nuclear] strike feasible, that would guarantee complete destruction of an enemy with no chance for retaliation. Maybe a technique for negating radar, for example. I can think of all kinds of possibilities. Would you want [that] loose in the world?"

There is a great deal of cynicism (or realism, if you're a cynic!) about how Washington would handle breakthrough information in the message concerning *non*-weapons technology as well. For example, if the message showed how to reprogram DNA to extend life then giving that knowledge to self-serving, ego-centric Washington hacks would simply mean that "we'll wind up with a bunch of immortal politicians, and nobody will ever hear of the technique again."

The paranoia isn't limited to the political world, but extends to the Church, too. As word of the message spreads, an American Cardinal gathers his staff together to discuss how to handle what he calls the coming of "a severe test of faith." The first issue on the table is one we've run into earlier in this book (back in Chap. 4), 'Do aliens have immortal souls?' When one of the staff wisecracks "Do we care?" another ignores the smart-aleck nature of the remark and simply repeats what I told you the Catholic theologian-reviewer in Sect. 4.1 wrote to me: "The ability to abstract from matter, to *think*, irrefutably defines the immortal soul." [To repeat what I wrote in Chap. 4: this is a *man-made* definition, and I see nothing in it that elevates it to "irrefutable" status.]

Well, then, with that 'settled' the gathering moves on to the Cardinal's fundamental concern: "What is the applicability of Christ's teachings to beings who are not born of Adam?" When that is at first dismissed as not being a serious issue, but rather as one that will concern only the "Bible-thumpers," the Cardinal disagrees. The problem as he sees it is that in the new message there are what appear to be schematic images of the aliens, images that are disturbing ones that "look like something out of Dali." One of the priests goes so far as to declare that "I'm certainly not prepared to believe that that odd little stick figure [in the message] is a picture of a creature with a soul." To that perhaps oddly non-Christian statement, the Cardinal points out that "if we can believe our experts, if we have indeed encountered aliens, whatever they

---

[28] It is, in fact, *thirty million* times the power-level of the most powerful commercial AM radio stations in America (50 kW = 0.05 MW).

look like, it will not be like us." That brings forth the observation that, of course, "the resemblance referred to in doctrine is of the soul, not of the body."

The Cardinal stuns his colleagues with this response: "Undoubtedly. But even so, we may find many among us who will be sorely tested by the notion of sharing salvation with large insects." He follows that rather graphic imagery with an even greater challenge to his listener's imaginations: "What would you say if their transmissions revealed them to be, by our standards, by the standards of the New Testament, utterly godless and amoral? Or worse, what if we are confronted by beings of compassion and apparent wisdom who, after a million years of examining the problem, have concluded that there is no God? Beings, perhaps, who have never even considered His existence?"[29] The group thinks about this, and then the Cardinal makes a politically crafty (if somewhat cowardly) decision: "We'll draft a letter to the pastors, to be kept in the strictest confidence. . . . Express our concerns. Instruct them, if questioned, to take the position that the revealed faith is God's message to man and has nothing to do with external agencies. Priests are not to bring the subject up."

Meanwhile, as the politicians in both Washington and Rome argue about what *might* be in the message, the scientists at Goddard continue to attempt to actually decode it. Eventually one of the scientists realizes that a sub-set of the message, a certain string of numbers, forms the description of a circuit schematic for some mysterious electronic device. When he builds it, however, and powers it up, nothing seems to happen. At least, not at first. After an hour of tinkering with the gadget, though, something odd *does* happen—he feels a "prickling" in his arm, and a near-by companion says "Something cold touched me!" And then his house burns down, killing him.

As the remains of the rear of the house smolder with heat, it is noticed that the front is super-*cold*. It seems, you see, that the alien device is a realization of the famous 'Maxwell's demon,' a microscopic creature imagined by the nineteenth century Scottish physicist Clerk Maxwell (1831–1879), who could separate 'fast' (i.e., *hot*) molecules from 'slow' (i.e., *cold*) molecules. This is a gross violation of what *we* know of the science of thermodynamics and McDevitt has imagined this 'impossible' device, one vastly different from the usual faster-than-light or time machine gadgets commonly found in SF stories, to illustrate just how far in advance the aliens are of Earthlings. It also gives some credibility to the 'hidden weapon' concerns.

---

[29] McDevitt almost surely wrote these words for his Cardinal with the inspiration of Blish's *A Case of Conscious*, published years earlier in 1958 (and which won the 1959 Hugo for Best Novel, one of the top writing awards in the world of science fiction). I'll discuss that important work in the next section, with special attention to how Blish handled the Cardinal's question.

The rest of the novel is a debate among the scientists on just what the message is all about—is it evil, or is it actually benign and so what appears to be evil is simply our inability to properly interpret it? McDevitt constructs all sorts of possibilities (and then has various story characters deconstruct them), but in the end he suggests what seems to be a God-free yet still 'religious' explanation. Yes, the universe is a hostile place, and yes there are *not* a lot of worlds with life. And so, rather than simply passively *listen* (as with the Hercules Project and Project Ozma), the aliens decided to be pro-active and *went looking* for 'others' with their artificially constructed Alpha/Beta system.

As the Goddard administrator explains to the physicist-priest Pete Wheeler, "We [have] insisted on perceiving them as a species like ourselves. But I think what we really have is a creature who is looking for something else alive and thinking in an empty universe. . . . All [of] those sterile worlds [in the universe]. Literally thousands of terrestrial planets, all embalmed in carbon dioxide or riddled with craters. It must be like that everywhere. And maybe, after we've advanced a little beyond where we are now, that emptiness will get to all of us . . ."

But it is with his next words that the administrator really stuns the priest: "the [aliens] are a group creature of some kind, a single intellectual entity. There's only one [alien]. It's damned near timeless. Immortal. And it's alone." That sounds like 'God' to me, but with a role-reversal that transfers the deliverance of salvation from the Biblical God to emanating from mankind itself.

6.4 Direct Encounters

Radio messages from the stars are fun, but of course nothing beats an actual *physical meeting* between humans and aliens. I've argued in this chapter that this is not likely, but nonetheless SF has eagerly embraced such meetings, and in this section I'll discuss just a few of the better stories. A prolific author of such stories was Stanislaw Lem, who wrote many short (usually hilarious) tales of interstellar adventure (collected in such anthologies as *The Star Diaries* and *Tales of Pirx the Pilot*). It was with his novels, however, that Lem seriously treated alien contact, sometimes with religious and moral commentary.[30] One striking feature of Lem's novels is how 'alien' are his aliens. One can understand why television's *Star Trek* had its aliens almost always 'appear human-like,' since real human actors had to play those roles, but SF writers are not so constrained. Many authors nevertheless fail to travel down the difficult road of creating really *alien* aliens.

---

[30] For an extensive discussion of this aspect of Lem's writing, see Kenneth Krabbenhoft's seminal essay "Lem as Moral Theologian," *Science-Fiction Studies*, July 1994, pp. 212–224.

Lem, however, never flinched. In *Eden* (1959) a starship crew encounters beings with large bodies that retract smaller torsos; in *Solaris* (1961) the entire ocean (!) of a planet is the lone sentient alien; in *Invincible* (1964) aliens appear as swarms of intelligent, self-reproducing micro-sized machines; and in *Fiasco* (1986) the aliens are what are mistakenly thought to be mere mounds in the ground. In this last novel, in particular, the influence of theology on Lem's writing is clear: the starship's crew includes a Papal envoy to the aliens (a Dominican priest) who is in conflict with the ship's Captain who, fears the priest, allows the ship's computer to make crucial decisions. The name of the computer is ironic, indeed: called the *Digitally Engrammic Universal System*, its acronym is DEUS (Latin for 'God').

A sub-genre of direct human-alien interaction fiction has humans dealing with alien *artifacts*, not with aliens themselves. Arthur C. Clarke was a master at such tales, with perhaps the 1951 story "The Sentinel" being the most famous. That tale (which was the inspiration for the film *2001: A Space Odyssey*) has human explorers on the moon triggering a mysterious monolith that they find there—a structure which, after being activated, sends a signal to (somewhere) with an alert that intelligent life on Earth has advanced to the first stage of space travel. More interesting for us, here, are Clarke's 1954 novel *Childhood's End* and the later (1973) novel *Rendezvous with Rama*.

In *Childhood's End*, aliens (known as 'the Overlords') exercise a benevolent but all-powerful control over human affairs. We have not gone 'there,' but rather they have come here. They have been 'on Earth' for 5 years when the novel opens (oddly, they have elected to remain out of sight and hover 50 km above the Earth's surface in huge spaceships). The Overlords have brought security, peace, and prosperity to the world, with all decisions at the international level made by them and yet, despite the resulting tranquility, the world's religions are in rebellion.

One of the Overlords explains to a human character why that is so: "You know why [religious men] fear [us]? . . . They know that we represent reason and science, and, however confident they may be in their beliefs, they fear that we will overthrow their gods. . . . Science can destroy religion by ignoring it as well as by disproving its tenets. No one ever demonstrated . . . the nonexistence of Zeus or Thor, but they have few followers now. [Religious men] fear, too, that we know the truth about the origins of their faiths. How long, they wonder, have we been observing humanity? Have we watched Mohammed begin the hegira,[31] or Moses giving the Jews their laws? Do we know all that is false in the stories they believe?"

---

[31] The flight of Mohammed from Mecca to Medina (the result of his belief in a single god, which put him in conflict with the polytheism of his time) in 622 A.D., an event marking the start of the Muslim era.

When asked the obvious question, "And *do* you?" the Overlord replies: "That . . . is the fear that torments them, even though they will never admit it openly. Believe me, it gives us no pleasure to destroy men's faiths, but *all* the world's religions cannot be right, and they know it. Sooner or later man has to learn the truth."[32]

This conversation, as have all the conversations between human and Overlord, takes place in a small room aboard one of the alien spacecraft, in which only the human is present. The Overlord's *voice* is heard by the human through a speaker. No Overlord has ever been seen, and nothing is known of their physical nature (other than they speak perfect English). By a clever subterfuge one human manages to get a momentary glimpse of an Overlord, but it is only later, when an Overlord finally reveals himself, that it becomes apparent why the aliens remained hidden: "The leathery wings, the little horns, the barbed tail—all were there. The most terrible of all legends had come to life out of the unknown past. Yet now it stood smiling, in ebon majesty, with the sunlight gleaming upon its tremendous body . . ." Apparently the Overlords were once spotted, long ago on a previous visit to Earth, and so was born the ancient tales of the Devil! (This is the one part of the novel that comes across as some sort of joke by Clarke.)

In *Rendezvous with Rama* an enormous, uninhabited, ten million megaton spacecraft suddenly appears and hurtles through the solar system. The novel is devoted to describing the physics of exploring such a gigantic structure (we are told it is a cylinder, 50 km long with a diameter of 20 km[33]), as well as speculation about its origin and purpose. One of the explorers is a member of the Fifth Church of Christ (he is a so-called "Cosmo Christer") which holds that Jesus was a visitor from the stars.

This explorer explains his interpretation of what is officially known as 'Rama' (after the Hindu god representing divine reasoning and virtue, although Clarke writes that it was a name chosen for no particular reason other than "long ago the

---

[32] Later in the novel we learn that the Overlords possess one of science fiction's classical gadgets, a so-called *time viewer*, which allows seeing the past, much like watching an old TV show. Able to look back as far as 5,000 years, the time viewer soon results in all of mankind's messiahs losing their divinity (no details are provided).

[33] The huge interior of this structure contains mountains, seas, and an atmosphere with weather, and the physics discussions are excellent. It is highly reminiscent of another masterpiece of astronomical construction (which may very well have inspired Clarke), the earlier (1970) novel *Ringworld* by Larry Niven. Ringworld, which is immensely larger than Rama, is also an alien artifact of mysterious origin. Built from a mass equal to that of Jupiter, Niven's ring has a radius of 93 million miles (Earth's orbital radius), is 1 million miles wide, and is 1,000 meters thick. Spinning around an axis (passing through a central star) normal to its plane at a speed of 770 miles per second, the apparent 'gravity' at the inner surface is very nearly one-gee. The habitable surface area is three million times greater than Earth's surface and so, even with *trillions* living on Ringworld, it would still feel almost empty. The physics of *Ringworld*, alas, is not as convincing as that of *Rendezvous with Rama*.

astronomers had exhausted Greek and Roman mythology; now they were working through the Hindu pantheon"): "Our faith has told us to expect such a visitation, though we do not know exactly what form it will take. The Bible gives hints. If this is not the Second Coming, it may be the Second Judgment; the story of Noah describes the first. I believe that Rama is a cosmic ark, sent here to save—those who are worthy of salvation."

Clarke doesn't do anything with this suggestion, though, and the novel ends with Rama exiting the solar system and continuing its enigmatic journey towards the Greater Magellanic Cloud. Rama has simply used its encounter with the Sun as a 'gravity sling-shot' to send itself on its way to some unknown goal. Clarke *does* make an interesting observation in the novel's final words, however, words that strongly hint at the miniscule importance of humans in the universe: "it had given [an] almost contemptuous proof of its total lack of interest in [a world] whose peace of mind it had so rudely disturbed." Rama was certainly no 'cosmic Ark' sent by God to save worthy Earthlings!

While in Lem's novels successful human communication with aliens is, at best, difficult (in *Solarius* it *never* occurs), other writers have explored the possibilities at the other extreme. That is, communication which is almost human-to-human. This does not mean there aren't problems! An interesting example of this is in Ray Bradbury's poetic 1951 short story "The Fire Balloons," in which a group of Episcopal Fathers fly to Mars on a rocket ship named *Crucifix* to save the souls of the human pioneer-settlers already on the planet. Sin on Mars, as the Fathers' Bishop puts it, "has collected there like bric-a-brac." The leader of the group, Father Peregrine, however, is eager to find some original, non-human Martians, too, who perhaps possess senses beyond the mere five of Earthlings. After all, he asserts, the more senses there are the greater the number of potentially interesting sins from which to be saved!

Father Peregrine is the author of a little book with the interesting title *The Problem of Sin on Other Worlds* (ignored, alas, by his Episcopal brethren as being "not serious enough"), and he has reached his curious conclusion by analogy with human senses and sins. As he enthusiastically explains to a skeptical colleague, "Adam *alone* did not sin. Add Eve and you add temptation. Add a second man and you make adultery possible. With the addition of sex or people, you add sin. If men were armless they could not strangle with their hands. You would not have that particular sin of murder. Add arms, and you add the possibility of a new violence. Amoebas cannot sin because they reproduce by fission. They do not covet wives or murder each other. Add sex to amoebas, add arms and legs, and you would have murder and adultery. Add an arm or leg or person, or take away each, and you add or subtract possible evil. On Mars, what if there are five new senses, organs, invisible limbs we can't conceive of—then mightn't there be five *new sins*?"

You can almost *hear* Father Peregrine rubbing his hands together in gleeful anticipation at the thought of all those delicious, new sins!

Once on Mars, Father Peregrine finds his original Martians. They aren't beings with bodies, though, but instead appear as floating, gaseous, fiery, blue spheres. As 'Fire Balloons.' At first he fails at every attempt to communicate with the Fire Balloons, and he begins to worry that they are *so* non-human that there could be no possible connection between them and God. Perhaps, in fact, there had been no Adam and Eve on Mars, and so no original sin, with the result that the Fire Balloons live in a state of grace. That is, they don't need salvation to keep their souls from the eternal damnation of Hell. Father Peregrine can't help but feel just a bit depressed at that—he could be out of a job!

At the end of the story, however, he finally achieves telepathic linkage with the aliens, and he learns that his theory of sin and senses is correct. What happened, long ago, is that the Fire Balloons did once have physical bodies, along with all the sins that come with them. But they learned how to free a man's soul and intellect from the body and so, now with none of the sins associated with the body to burden them, they live in God's grace. In Bradbury's tale, God has not limited His presence to Earth, and the story ends on this happy note: "There's a Truth on every planet. All [are] parts of the Big Truth . . . We'll go on to other worlds, adding the sum of the parts of the Truth [from each new world] until one day the whole Truth will [be known]."

Far less happy are the novels of Blish (*A Case of Conscious*) and Russell (*The Sparrow* and *Children of God*), in which direct human contact with aliens has disastrous consequences. In each the central protagonist is a Jesuit priest-scientist whose very faith is shaken to the core. I'll start with *A Case of Conscious*.

Father Ramon Ruiz-Sanchez is a biologist, a Peruvian Jesuit priest, and part of a four-man advance evaluation team on the planet Lithia, 50 light-years from Rome. The Team's mission is to decide whether or not the remarkably Earth-like planet, home to an intelligent race of 12-ft tall reptiles, can be a useful port-of-call for Earth, without risking either humans or Lithians. (If that sounds a lot like the *Star-Trek* Federation's well-known forbidding of any damaging first-contacts with newly discovered life-forms—the so-called 'prime directive'—that's because Blish was a major contributor to the television show.) To get the Team to Lithia, Blish has to somehow overcome the vastness of that 50 light-years, and he does that with what can only be called SFBS: "highly esoteric tampering with the Haertel equations—that description of the space-time continuum which, by swallowing up the Lorentz-Fitzgerald contraction exactly as Einstein had swallowed Newton (that is, alive), had made interstellar flight possible. Ruiz-Sanchez did not understand a word of it, but, he reflected with amusement, it was doubtless perfectly simple once you understood it."

Father Ruiz-Sanchez is a deeply religious man who finds no difficulty in accepting the reality of intelligent life beyond Earth; as he tells one of the other members of the Team (Cleaver, an atheist physicist), "For me, biology *is* an act of religion, because I know that all creatures are God's—each new planet, with all its manifestations, is an affirmation of God's power." But there *is* a 'problem' with the Lithians, themselves. It is, perhaps ironically, that there *are* no problems, none at all. Their world is *perfect*. The Lithians have no crime (the concept of a locked door is a mystery to them), no cults, no separate nations at odds with each other. The entire planet is a homogeneous whole, with all Lithians speaking a single language, with never a harsh word uttered. Lithian society seems to be one in which only saints exist.

It seems, in fact, too good to be true and, wonders the priest, perhaps it *is* too good, as the Lithians also have no religion and so no concept of God. Lithia is a Garden of Eden before the Fall of Adam and Eve, inhabited by intelligent, supremely rational beings with tails, beings that are more like thinking machines. They are creatures lacking nothing but souls to be saved. As Father Ruiz-Sanchez explains his quandary to the other members of the Team, "Here on Lithia, fifty light-years away from earth and among a race as unlike man as man is unlike the kangaroos, what do we find? A Christian people, lacking nothing but the specific proper names and the symbolic appurtenances of Christianity. I don't know how you three react to this, but I find it extraordinary and indeed completely impossible—mathematically impossible—under any assumption but one."

Father Ruiz-Sanchez then shocks the Team by stating that assumption.

"We have," he says, "a planet and a people [created] by the Ultimate Enemy. It is a gigantic trap prepared for all of us—for every man on Earth and off it. We can do nothing with it but reject it, nothing but say to it, *Retro me, Sathanas*. If we compromise with it in any way, we are damned." For Father Ruiz-Sanchez, Lithia is a creation of Satan, a planet designed by Evil to lead all who visit it to come to believe that spiritual perfection *without* God is possible. And this is why his vote is "to seal Lithia off from all contact with the human race. Not only now, or for the next century—but forever." This is a dangerous position for the priest to take because, as one of the Team observes, "To set such a trap, you must allow your Adversary to be creative. Isn't that—a heresy, Ramon?"

The physicist, Cleaver, views Lithia differently. The planet's very name, inspired by the abundance of the element lithium which is crucial to the construction of nuclear weapons, tells us what he is fascinated by—he sees the planet as a virtual 'cornucopia of hydrogen fusion bombs'! His vote is to *not* forbid access to Lithia, but rather to treat it as a vast munitions arsenal. The other Team Members are uncertain, and so the vote splits.[34] Its job done, the Team returns to Earth.

But they don't return alone. As a parting gift, an embryonic Lithian child is given to the Team, to be raised on Earth as literally a 'stranger in a strange land.' This proves to be a disaster, however, as the child grows to be an adult who is both ignorant of Lithia and repulsed by Earth's society. Meanwhile, after years of pondering, the Pope summons Father Ruiz-Sanchez to Rome, a summons the priest fears means he is at last to finally be charged with the heresy of proclaiming Satan to be creative. This is not the case, however, and instead the Pope has at last become convinced that Father Ruiz-Sanchez is correct and that there is but one choice on what to do about Lithia: *the entire planet must be exorcised.*

As the Pope explains to the stunned priest, Satan has no power to create, only to deceive, and so all that the Team 'saw' during its visit must have been simply a massive hallucination. This might seem to be a decision that validates the priest's vote, but now Father Ruiz-Sanchez is not so sure of things, even if it is "easier to believe in a planet-wide hallucination ... than in the heresy of satanic creativity." His new concern is fired by the fact that the rebellious Lithian child, now full-grown, has returned to Lithia to foster rejection of the planet's perfection, and Father Ruiz-Sanchez now worries from the opposite extreme: "What if he were wrong after all? Suppose, just suppose, that Lithia were Eden, and that the Earth-bred Lithian who had just returned there were the Serpent foreordained for it?" Would he be destroying God's version of the Fall on Lithia?

Nevertheless, obeying the order of the Holy Father, the exorcism takes place from an observatory constructed in a crater on the moon. The observatory is equipped with a telescope and a viewing screen that has the fantastic property of imaging Lithia, 50 light-years distant, so that it appears to be only 250,000 miles away. That is, as close as is Earth. Not only that, the telescope tosses Einstein's insights on simultaneity out of the window, as we are told (by a stereotypical technical geek) "We have spanned not only the space, but also the time ... What we are seeing [on the viewing screen] is Lithia *today* ... not Lithia fifty years ago." To that astonishing claim comes what has to be the most enormous understatement in the history of science: "Congratulations."

Father Ruiz-Sanchez then delivers the ritual words of exorcism, ending with the thunderous line "I SAY UNTO YOU, ANGEL OF PERDITION: DEPART, DEPART, DEPART!" That works, and the whole, perfect planet of Lithia first swells like a balloon and then vanishes in a brilliant blue-white

---

[34] Some critics have speculated that Cleaver's name was chosen by Blish precisely because of his role in splitting the vote.

glare. The telescope electronics goes dead, as well "with a puff of fuses" (just to drive home how stupendously energetic is the departure of Satan). Still, while the priest was clearly correct in his suspicions, he nonetheless feels a great loss, too, as the last line in the novel reveals: "When Father Ramon Ruiz-Sanchez . . . could see again, they had left him alone with his God and his grief."

Far different from Blish's work are the two novels by Russell, *The Sparrow* and its sequel *Children of God*. Both are, I think, brilliant examples of character construction, and each represents great knowledge by Russell of both the scientific life[35] and of the Church. Both are, fundamentally, studies in the loss of faith of a priest who loves God but who eventually comes to believe (because of what happens to his friends and himself on an alien planet) that God has played a terrible joke. In addition, the interaction between human and an alien culture is carefully developed, and the disastrous consequences seem to be inevitable. There are no supernatural aspects in either novel, no appearances God and/or Satan, and only a small amount of straight SF (but what there is done well—the non-intuitive temporal physics of near light-speed interstellar travel is nicely *and correctly* presented[36]).

Most of the events in *The Sparrow* take place in one of two time periods. The first one is just before the reception of radio signals from space at the Arecibo Radio Telescope facility in Puerto Rico (in 2019). Soon after, the launch of an interstellar space ship by the Society of Jesus occurs, with a crew of eight, to a planet near Alpha Centauri, 4 light years distant. The second time period describes the reception back on Earth (in 2060) of the lone survivor of that trip, Father Emilio Sandoz who is a skilled linguist. The novel moves back-and-forth between those two periods, and what at first appear to be mysterious events slowly have their explanations revealed. (There is a third time period as well; the 4 years prior to the first period, in which the character development of some of the central story figures takes place, but that period isn't crucial for my remarks here.)

The story line is a simple one: radio signals received at Arecibo reveal that there is intelligent life on a planet orbiting the star nearest to Earth; the Society of Jesus finances a journey to that planet, in a spaceship made from an old space-mining asteroid which is described as a 'rock that looks like a giant potato'—the crew lives in the asteroid's hollow core with the outer layers of rock serving as shielding from the effects of moving at high speed through

---

[35] Mary Doria Russell (born 1950) is not a professional SF writer. Rather, she is a professional writer who has written two terrific SF novels. She has a doctorate in biological anthropology.

[36] According to Einstein's special theory of relativity, the rate of a clock moving relative to a stationary clock (a spaceship clock and a clock on Earth, respectively) are different. The stationary clock runs faster than does the moving clock. So, an interstellar journey at near light speed will take a long time in Earth years, but not so long in space ship years. See note 3 again.

space dust; the trip is made by constantly accelerating at one-gee until halfway there, and then decelerating at one-gee for the second half; after a journey lasting 6 months (in ship time) and 17 years in Earth time,[37] the asteroid arrives at and goes into orbit around its destination, where the crew finds there are *two* sentient species on the planet; the two species are physically nearly the same (very large, kangaroo-like beings), with one of them in a subservient role to the other, much like humans would treat very intelligent dogs; the dominant species periodically slaughters the other for meat and population control; the humans are so horrified at this that they cause rebellion and wide-spread bloodshed between the two species; all the humans are dead by the end of the rebellion, with the exception of Father Sandoz who is held captive and subjected to physical mutilation of his hands and to periodic rape (I found this to be the most difficult part of the novel to accept, that aliens would find a human sexually provocative[38]); a second space mission (sent to find why all connection has been lost to the first) frees Father Sandoz from captivity, but only after he, to the horror of his rescuers, murders a young female of the subservient species who has led them to Father Sandoz; his rescuers place Father Sandoz back on the asteroid spaceship and send it by auto-pilot back to Earth (arriving in the year 2060), where he is immediately put into seclusion by the Society of Jesus.

The rest of the novel is the slow unraveling of what has happened to Father Sandoz, who at the start of the novel is a happy, joking, free-wheeling man but who, upon his return to Earth, is a bitter, physical and spiritually broken, nearly unrecognizable shell. The answer is that Father Sandoz's faith has been almost entirely destroyed. He cannot reconcile what he initially thought was God's wish for the mission to occur, with the horrible sequence of events that then occurred. As Father Sandoz notices how all of the multitude of necessary conditions for the trip to Alpha Centauri become satisfied, "it became hard to ignore how, against odds, the dice kept coming up in favor of the mission." So, why did it all so tragically fail?

The Father General of the Society of Jesus, an old acquaintance of Father Sandoz, is sympathetic to his friend's plight and tells him during the seclusion

---

[37] When Father Sandoz, who you'll recall is a linguist and not a physicist, has the slowing of ship's time relative to Earth's time explained to him, he quite naturally asks "Why does it work that way?" One of the other Jesuits in the crew gives him this answer, one that might appeal to readers who aren't so enamored with mathematical physics: *"Deus vult, mes amis"* ("God likes it that way.") More historically, *Deus vult* ("God wills it") was the battle cry of Christian warriors in the First Crusade against the Muslims to recover the Holy Land in the eleventh century.

[38] As one reviewer bluntly stated this issue on the Web: "[V]ery probably none will have known anyone who was buggered by an intelligent flesh-eating kangaroo. If one did, one could presume to wish it hadn't happened . . . [I]n the climate of much of the English-speaking world, a story in which a priest is on the receiving end could easily be told as a bad joke . . ."

"Emilio, everything I have learned about the mission leads me to believe that you went for the greater glory of God. You believed that you and your companions were brought together by the will of God and that you arrived at your destination by the grace of God." And even after his imprisonment by the aliens, Father Sandoz tells the Father General that "I believed that God was with me." He *did* feel he was in God's hands, and that "whatever happens now to me is God's will."

And then, finally, Father Sandoz reveals the horror[39] of what did happen: "I was raped." As he explains to his interrogators on Earth, "You see, that is my dilemma. Because if I was led by God to love God, step by step, as it seemed, if I accept that the beauty and the rapture were real and true, then the rest of it was God's will, too, and that, gentlemen, is cause for bitterness. . . . If, however, I choose to believe that God is *vicious*, then at least I have the solace of hating God." Interestingly, Father Sandoz does not blame Satan for his torture because "Satan ruins people by tempting them to take an easy or pleasurable path," which certainly wasn't what happened to him.

The Father General at last understands what has happened to Father Sandoz, when he says "What a wilderness, to believe you have been seduced and raped by God." Still, while understanding Father Sandoz's despair, the Father General continues to believe that God still passionately cares about all humans, and he quotes Matthew 10:29: "Not one sparrow can fall to the ground without your Father knowing it"—and so we at last see where Russell got the novel's title. He is brought-up short, however, when reminded that even if God knows of the sparrow's fall, nonetheless the sparrow still falls.

The novel ends with Father Sandoz still deep in depression. When he is told that the Society of Jesus is sending another mission back to Alpha Centauri, and that it is desired that he return to the scene of so much anguish ("We could use your help. With the languages."), he flatly refuses to consider it, even after learning that one of his fellow missionaries on the first mission survived the rebellion. And that is where the sequel novel, *Children of God*, picks-up the story.

Father Sandoz's physical condition has greatly improved as the novel opens, but his emotional and spiritual destruction is apparently permanent. Indeed, not only does he continue to refuse to return to Alpha Centauri, he has decided to leave the Society of Jesus and to marry. These decisions are aborted, however, when he is taken prisoner and forced to join the return mission. Arriving back at Alpha Centauri, the two sentient races are at war, a

---

[39] The 'murder' of which Father Sandoz stands accused was committed in his despair at imprisonment and so, despite his vows as a priest, he vows to kill the first of his alien tormentors to next enter his cell. Unfortunately, it was his alien rescuer who was that next alien, and he killed her before realizing his mistake.

continuation of the rebellion instigated by the other survivor of the first mission. The entire novel is then devoted to the cultural upheaval that first contact has initiated, with many tears and much bloodshed and death on nearly every page. Unlike many religious first contact stories, *Children of God* features neither God nor Satan (outside of philosophical arguments[40]), or the extremes of showcasing human brilliance or ignorance. It is simply a tale of people doing the best they can in difficult circumstances, and often making a mess of it all.

To end on a somewhat dark note, you might ponder the inversion of the stories considered so far, all of which presume that humans, as they spread out into the universe to preach the word of God, will make religious contact with aliens *on their worlds*. What if, instead, alien priests came *here* to spread the word of *their* god(s)? What then? One grim answer is provided by "In His Own Image" (Payes), which ends with these words from such a visitor, just after it has destroyed a church of a "false god" depicted by a Cross bearing a crucified body: "[It] vowed to spread the True Faith over all this alien, hateful planet called Earth." The destroyed Cross is then replaced by an enormous wheel bearing the dead body of a hideous creature with faceted eyes, antennae, and six appendages that each end in a great claw. Which, of course, could still have been Jesus 'in the image' of the alien world he visited. As Father Peregrine asks, in Bradbury's "Fire Balloons," "If Christ had come to us on Earth as an octopus, would we have accepted him readily?"

Well, perhaps that is just a bit *too* dark for our ending, so let the following question be our conclusion: From where comes the need, in both human and alien missionaries, to spread the word of God throughout the science fictional universe? Perhaps one good answer comes from the Father General of the Society of Jesus, in *The Sparrow*. When thinking of a seventeenth century French Jesuit who endured extraordinary hardship and, finally, an agonizing death, when preaching in the New World to Indians, he decides it wasn't madness that drove that priest, or any of his colleagues, but rather it was "the mathematics of eternity that drove them. To save souls from perpetual torment and estrangement from God, to bring souls to imperishable joy and nearness to God, no burden was too heavy, no price too steep." SF missionaries, both human and alien, share that belief—but it is a belief that brings with it great risk. As the Father General thinks, in *Children of God*, "First contact—by definition—takes place in a state of radical ignorance." That ignorance makes catastrophic disaster a virtual certainty.

Along with the risk comes a personal benefit, however, one that any honest missionary would surely have to admit borders dangerously close to the sins of

---

[40] Just one example: "I've always thought it was a tactical mistake for God to love us in the aggregate, when Satan is willing to make a special effort to seduce each of us separately."

pride, curiosity and/or ambition. I am referring to what is nicely described in the opening to Shelley's *Frankenstein*, in the first letter the sailor Robert Walton writes from St. Petersburg, Russia to his sister in England. That novel is mostly Walton's description of the story related to him by Doctor Frankenstein (who has been rescued, when nearly dead, by Walton's ship in the Arctic) while in pursuit of his monster. The first letter is before all that, however, and is simply the *reason d'état* for why Walton is searching for the legendary, long-sought Northwest Passage. His words to his sister are, I think, just what an honest interstellar missionary would also write to explain *his* motivation:

> "I shall satiate my ardent curiosity with the sight of a part of the world never before visited, and may tread a land never before imprinted by the foot of man. These are my enticements, and they are sufficient to conquer all fear of danger or death and to induce me to commence this laborious voyage with the joy a child feels when he embarks in a little boat, with his holiday mates, on an expedition of discovery up his native river."

To end this chapter with a *personal* comment about humankind's place in the Universe, and on whether we have a special role in God's plan, consider these beautiful words, the rarely sung second verse from the otherwise well-known American song *Home on the Range*, written in 1873:

> How often at night when the heavens are bright
> With the light from the glittering stars,
> Have I stood here amazed and asked as I gazed
> If their glory exceeds that of ours?

This mysterious question, 'are we God's special creation?,' is clearly one that has puzzled all who have pondered it. It certainly long pre-dates today's SF.

If 'they' *are* 'out there' then the answer is almost surely titled towards YES, if we remember that "we are living on an insignificant speck of rock going around an undistinguished star in a low-rent section of the galaxy"[41] in just one of a hundred billion galaxies. This fact does prompt again the question I asked at the start of this chapter: why did God choose to place His chosen on the cosmic equivalent of a tiny island, one lost in the vast ocean of space? I expect some will say 'To keep us humble,' but I think that just a bit weak, the equivalent of what I think the equally weak 'He works in mysterious ways.'

---

[41] The quotation is from Robert T. Rood and James S. Trefil, *Are We Alone? The Possibility of Extraterrestrial Civilizations*, Scribner's 1981. A 1995 book (that I highly recommend) with the same title, dealing with the philosophical implications of the discovery of extraterrestrial life, was written by the British theoretical physicist Paul Davies. Davies received the 1995 Templeton Prize "for contributions affirming life's spiritual dimension."

Chapter 7
Time Traveling to Jesus

7.1 Time Travel: Fact or Fantasy?

Science fiction stories have long been filled with marvelous gadgets, some of which have appeared in earlier chapters. Just to mention a few, even if I repeat myself, they include automatic language translators, invisibility cloaks, faster-than-light space ships slipping in and out of something called hyper-space, supercomputer brains, matter transmitters ('beam me up, Scotty'), disintegrator ray guns, and tractor beams (think of a ray that *pulls stuff towards you*). But the ultimate SF gadget has to be the time machine. The other gadgets do neat things, sure, but that's pretty much the extent of them. Time machines, while a gadget that opens the future to us, also appeals to our nostalgia for the past.

As the English writer Virginia Woolf (1882–1941) wrote shortly before her death,[1] "Is it not possible—I often wonder—that things we have felt with great intensity have an existence independent of our minds; are in fact still in existence? And if so, will it not be possible, in time, that some device will be invented by which we can tap them? ... Instead of remembering here a scene and there a sound, I shall fit a plug into the wall; and listen in to the past. I shall turn up August 1890." Besides the nostalgia, however, the potential paradoxes that are intimately linked to time travel to the past challenge our brains in ways that the other gadgets do not. Time travel makes you *think*, but often you'll experience the feeling that a time traveler has with paradoxes in Robert Heinlein's 1941 story "By His Bootstraps": "He felt the intellectual desperation of any honest philosopher. He knew that he had about as much

---

[1] In her essay "A Sketch of the Past," *Moments of Being* (Jeanne Schulkind, editor), The University Press 1976, pp. 67. "Putting a plug into the wall" and listening to the past sounds a lot like today's internet. After all, isn't that what happens when you stick your computer's high-speed internet cable into the jack on the wall and watch History Channel videos of World War 2?

P.J. Nahin, *Holy Sci-Fi!*, Science and Fiction,
DOI 10.1007/978-1-4939-0618-5_7, © Springer Science+Business Media New York 2014

chance of understanding such problems as a collie has of understanding how dog food gets into cans."

Here's an example of what Heinlein meant. A 1953 story by Charles Harness (1915–2005), "Child by Chronos," begins with the birth in 1957 of a girl. After 20 years of intense competition with her mother (who has an uncanny ability to predict the future), she travels back in time from 1977 to a few months before her own birth. She becomes pregnant (by a man she later discovers is her father!) and gives birth to a girl. The new mother has, of course, knowledge of all that will happen during the next 20 years, including the fact that she will have an intense completion with her rebellious daughter . . . This probably appears to you to be on shaky biologically ground because a child gets only half its genes from each parent, and so a daughter should be only *half* what her mother is, and not identical (they are, of course, the same person). On the other hand, both mother and daughter do have the same parents and so, perhaps, it isn't impossible that they would have the same genetic description. Well, whatever you may think of all this, my point is that you *are thinking*![2]

The logical paradoxes, at least at first thought, seem to be unanswerable, with the famous 'grandfather paradox' the best known (it is only one of several distinct logical paradoxes[3]). After all, what sense can there be to any attempt by a time traveler to go back into the past to kill his grandfather when that ancestor is still a baby? Or to put it even more directly, can a time traveler to the past, in a form of high-tech suicide, kill *himself* as a baby? How then could the time traveler come to be born? The classic use of the grandfather paradox is as a (false) proof that time travel to the past is impossible.

Certainly the time paradoxes were just too much for Isaac Asimov to accept. As he once put it, "The dead give-away that time travel is flatly impossible arises from the well-known 'paradoxes' it entails . . . So complex and hopeless are the paradoxes . . . so wholesale is the annihilation of any reasonable concept of causality, that the easiest way out of the irrational chaos that results is to suppose that true time travel is, and forever will be, impossible."[4]

It's curious that Asimov wrote those words in 1984 because, years earlier in 1949, one of the twentieth century's great mathematicians, Kurt Gödel

---

[2] As bizarre as Harness' story may seem, Heinlein's famous 1959 time travel twister "All You Zombies—" is even more convoluted, with the tale of a person who is his/her own father *and* mother! As one wit once put it, it is tale that demands to be told in the first person (which is precisely how Heinlein actually wrote it).

[3] Numerous time travel paradoxes, and some of their possible resolutions, are treated at length in my book *Time Machines: time travel in physics, metaphysics, and science fiction*, Springer-AIP 1999, pp. 245–353.

[4] See Asimov's editorial "Time-Travel" in *Asimov's Science Fiction Magazine*, April 1984. His negative view of time travel didn't prevent Asimov from using it in many of his stories, however, often to great effect.

(1906–1978), had shown that Einstein's equations of general relativity contained solutions that permit time travel to the past.[5] With the publication of Gödel's paper (which Asimov apparently missed) time travel to the past moved, overnight, from the realm of fantasy SF to respectable theoretical physics. Gödel, himself, appears to have seriously thought time travel to the past could be the explanation for events that have been the cause of reports over the centuries of 'ghosts.'[6]

Since Gödel's pioneering paper, physicists have discovered several different *theoretical* time machines: the wormhole time machine of Kip Thorne (Caltech), the rotating cylinder time machine of Frank Tipler (Tulane), and the cosmic string time machine of Richard Gott (Princeton). All of these mechanisms require fantastic amounts of mass-energy to function, and none are presently being put together (as far as I know) in anyone's basement. But the very fact that they are even theoretically possible (unlike a perpetual motion machine) is all that is needed to make the time machine a 'plausible' *science fiction* gadget[7] and not just a fantasy plaything.

Now, most certainly not all physicists are willing to entertain the possibility of time travel to the past, even in the face of the above developments. The most famous of the doubters is the English theoretician Stephen Hawking (born 1942), who thinks it is all just hogwash (and he very well might be right). Nevertheless, he too studies time travel because, as he admits, there is nothing in *known* physics that specifically forbids it and so his goal is to discover the new physics he is sure exists that *will* forbid time travel. For now, his only argument for his position is what he tongue-in-cheek calls the *Chronology Protection Conjecture*[8]: time travel to the past is impossible, thus "making the universe safe for historians" who otherwise would have to worry about those who try to alter the past for their own gain because they believe it is possible to 'change yesterday today for a better tomorrow.' Hawking was actually anticipated in his Conjecture by SF writer Larry Niven, who had specifically put it forth as *Niven's Law*, 20 years earlier: "If the universe of discourse permits the possibility of time travel, and of changing the past, then no time machine will be invented in that universe."[9]

---

[5] See Gödel's paper "An Example of a New Type of Cosmological Solutions of Einstein's Field Equations of Gravitation," *Reviews of Modern Physics*, July 21, 1949, pp. 447–450.

[6] See *Biographical Memoirs of Fellows of the Royal Society*, November 1980, pp. 148–224.

[7] The *physics* of all of these various time machines is discussed in my *Time Machines*. See also Bud Foote, *The Connecticut Yankee in the Twentieth Century*, Greenwood Press 1991.

[8] See Hawking's famous paper, "Chronology Protection Conjecture," *Physical Review D*, July 1992, pp. 603–611. Hawking's Conjecture is an obvious echo of the Fermi Paradox.

[9] See Niven's essay "The Theory and Practice of Time Travel," in *All the Myriad Ways*, Del Ray/Ballantine 1971.

This perhaps cryptic sentence might be clearer if put this way—in a world in which time travel to the past is possible and which allows a time traveler to make changes, events after the change is made will be altered and this includes the events which resulted in the time journey in the first place. Thus, the time journey itself will be altered, and this includes the nature of the change in the past. This rippling-through-time process continues until *eventually* (whatever that word might now mean) a steady-state historical reality is reached in which no further adjustments occur. That is, a world in which *no* time journey occurs! A clever novel that develops this idea is the 1974 *Times Without Number* by John Brunner, about which I'll say more in just a bit.

Hawking has no formal proof of his Conjecture (that's why it's a *conjecture!*), other than the observation that the present hasn't been overrun by vast hordes of tourists from the future. This is amusing, yes, but it is actually a pretty thin argument because all of the theoretical time machines I mentioned earlier share the common property of not being able to visit the past more remote than the date of the machine's creation. (It is also amusing to note that this property was not first mentioned in a physics journal, but rather in a science fiction story published nearly 80 years ago.[10]) All that Hawking's observation shows is that *if* time machines are possible *then* one hasn't yet been constructed.

Even before Niven, Robert Silverberg had essentially stated Hawking's Conjecture in the form of the so-called *cumulative audience paradox*, in his 1969 novel *Up the Line*. That paradox claims that as time travelers to the past continue to visit certain historically interesting dates and places, there will be an ever-increasing number of people present. As stated in the novel, "Taken to its ultimate, the cumulative audience paradox yields us the picture of an audience of billions of time travelers piled up in the past to witness the Crucifixion, filling all the Holy Land and spreading out into Turkey, into Arabia, even to India and Iran . . . Yet at the original occurrence of [that event] *no such hordes were present*." (The modern view of this is that the use of the word *original* is misleading, in that it implies the past happens more than once.) Was it Silverberg's SF novel, perhaps, that was the inspiration for Hawking's 'no hordes of tourists from the future' observation that he claims as 'experimental evidence' for the Chronology Protection Conjecture?

If it wasn't Silverberg or Niven, however, then perhaps *another* SF author, Arthur C. Clarke, was Hawking's inspiration. In a 1985 essay he wrote "The most convincing argument against time travel is the remarkable scarcity of time travelers. However unpleasant our age may appear to the future, surely

---

[10] See "The Time Bender" by Oliver Saari (1919-2000), which appeared in the August 1937 issue of *Astounding Stories*.

one would expect scholars and students to visit us, if such a thing were possible at all."[11] And another SF writer, Jack McDevitt, writing 2 years before Hawking's Chronology paper appeared, had a character in his 1989 short story "Time's Arrow" say "If [time travel] *could* be done, someone will eventually learn how. If that happens, history would be littered with tourists. They'd be *everywhere*. They'd be on the *Santa Maria*, they'd be at Appomattox with Polaroids, they'd be waiting outside the tomb, for God's sake, on Easter morning." (But not, as I said earlier, if no time machine has yet been constructed.)

7.2 Theology and the Past

The now theoretically possible time machine offers the tantalizing possibility for 'realistically' indulging our imaginations in one of the oldest of fantasies—the changing the past. When I say it's an old idea, I mean that it can be traced back to at least four centuries before Christ. That's when Aristotle, in his *Nicomachean Ethics*, wrote that the Greek poet Agathon, a century earlier, had declared "For even God lacks this one thing alone. To make a deed that has been done undone." That is, as Ralph Waldo Emerson wrote in his poem "The Past," not even God can manipulate the past: "All is now secure and fast, Not the gods can shake the past."

The author of "The Mosaic," which appeared in the July 1940 issue of *Astounding Science Fiction*, didn't agree with Emerson. In that story the Moslem defeat by Christians in 732 'originally' is a Moslem victory. Centuries later the first time traveler (a Moslem) accidently changes the victory into a defeat and one of the repercussions is that *he* will never be born. So, he vanishes "with all the suddenness of a bursting bubble. And with him into nothingness, across the gulf of Time" goes all the 'original' history after 732, having been changed to 'our' world's history that records the ancient victory of Cross over Crescent.

The rigidity (or not) of the past is of special interest to theologians because it is directly related to the question of free-will versus fatalism. That is, are humans the creators of the future, or are they mere fated puppets of destiny? Is a time traveler to the past unable to alter events because that was the *only* way they *could* happen? The Bible, alas, offers no definitive help on answering such questions. In his *Guidance to the Duties of the Heart*, the Jewish eleventh-century Spanish rabbi and philosopher Bahya ibn Paquda lists several

---

[11] See Clarke's essay "About Time" in *Profiles of the Future*, Warner 1985.

scriptural texts in support of predestination, and yet he also offers another Biblical list in support of free-will. For example, compare *Pslam* 127 with *Job* 34:11. *Guidance* aptly presents its lists in the form of a dialogue between the (rational) mind and the (emotional) soul. In this dialogue the mind attempts to ease the soul of its concern with the "ills of the body," one of which is the conflict between free-will and fatalism.

The Italian cleric Peter Damian (1007–1072/3), who became a Christian saint, had a slightly different take on this issue. He believed that *nothing* could withstand the power of God, not even the solidity of the past. Writing in his *De Omnipotentia Dei* ("On the Divine Omnipotence in Remaking What Has Been Destroyed and in Undoing What Has Been Done"),[12] he declared "Just as we can duly say "God was able to make it so [that] Rome, before it had been founded, should not have been founded,' in the same way we can equally and suitably say, 'God can make it so that Rome, even after it was founded, should not have been founded.'"

Two centuries later Aquinas argued the opposite position, that changing the past is *not* within the power of God. Whereas Damian felt it impossible to deny any act to God, Aquinas took the far more moderate position that part of God's law is that there be no contradictions in the world and that certainly God would be bound by his own law. (You'll recall that Arthur C. Clarke took that position in his essay "God and Einstein" that I mentioned in Chap. 1, Sect. 1.4.) As Aquinas argued, "It is best to say that what involves contradiction cannot be done rather than God cannot do it."

Aquinas' point, that of avoiding contradiction, is central to the modern understanding of time travel. To make sense to physicists, time travel to the past must *always* be self-consistent. That is, none of the events along a time traveler's trajectory through spacetime can be in conflict. This requirement has, in fact, been elevated to the level of an axiom called *Novikov's Principle of Self-Consistency* (after the Russian theoretician Igor Novikov, born in 1935). This principle is basic to understanding the difference between *changing* the past and *affecting* (or influencing) the past. If you journey back to view the Great Fire of London in 1666 it is logically impossible for you to prevent the fire (and so change the past) but it *is* logically possible for you to affect the past if you were the one who started the fire. In the same way, you might be

---

[12] This work was written as letter to his friend Desiderius (who became Pope Victor III), in which Damian rebutted Desiderius' defense of St. Jerome's claim that "while God can do all things, he cannot cause a virgin to be restored after she has fallen." Desiderius thought the reason God could not restore virgins is that he does not want to, to which Damian replied that this would mean God is unable to do whatever he does not want to do, and *that* would mean that God would then be less powerful than mere men, who *are* able to do things they don't want to do (such as go without food for a month). This is a good example of the risk you run by getting into a debate with a theologian! (Remember *my* linguistic encounter, over the issue of 'who has a soul?', with a theologian?)

the one who threw a match on the faggots at the feet of Joan of Arc, but your time trip will be in vain if you hope to put the flames out with a fire extinguisher.

In his 1949 story "The Biography Project" Horace Gold (1914–1996), the editor of *Galaxy Science Fiction Magazine*, cleverly illustrated the distinction between affecting and changing the past. The wonderful SF gadget in Gold's tale isn't a time machine, but rather its first cousin, the so-called *time viewer* (called the Biotime Camera in the story), which can film (alas, however, without sound) the past. Using this gadget, the Biofilm Institute funds teams of biographers to study the lives of past notable personages. In particular, the lives of those who developed neurotic psychoses, such as Marcel Proust and Isaac Newton. And, indeed, the Biotime Camera does capture images of these individuals as they begin to display increasingly disturbed behavior.

We see Newton, for example, begin to peer into dark corners, looking for those he has come to believe are spying on him. On his death bed, the biography team assigned to him reads his lips and discovers that his final words are "My guardian angel. You've watched over me all my life. I am content to meet you now." It is then that the Biofilm Institute realizes what it has done. Newton *was* in fact being spied upon—by the Biotime Camera, which has not changed the past but *has* affected it. (This is, of course, really just a stimulating exercise in speculative fiction, as present medical thought is that Newton's odd behavior was actually due to mercury poisoning from his alchemy experiments and not from being time-viewed! Still, it's something to keep in mind when you next do something you'd rather not have appear in a fortieth century doctoral dissertation about *your* life!) Gold's story may have had a curious theological result in the real world, as shortly after it appeared an Italian Benedictine Monk, Pellegrino Maria Ernetti (1925–1994), announced his claim to have used what he called a *chronovisor* to, among other things, photograph Jesus as he died on the Cross.[13]

Here's an elementary SF theological example of *affecting* (but not changing) the past; in the 1949 story "Uncommon Castaway" by Nelson Bond (1908–2006), time travelers journey back to Biblical times in a submarine equipped with a time machine. While sailing in an ancient sea the crew comes across a man who has been set adrift, and so rescue him. After their return to the present they realize that the man they took aboard was Jonah—and so *that*

---

[13] An odd little book that describes this aspect of Father Ernetti's life is by Peter Krassa, *Father Ernetti's Chronovisor: the creation and disappearance of the world's first time machine*, New Paradigm Books 2000. Krassa's book will appeal most strongly to the highly gullible.

explains the famous 'swallowed by a whale' tale! The time journey didn't change the past, but it certainly affected it.

Just 3 years later a far more sophisticated 'affecting the past' time travel tale appeared, one echoing the ministry of Jesus. In the middle of the twenty-second century, in the 1952 story "The Skull" by Philip K. Dick, a man named Conger, in prison for illegal hunting, is offered a curious deal. His sentence will be cancelled if he agrees to travel 200 years back in time to hunt down and kill the Founder of a religious Movement called the First Church. Conger's hunting skills are of particular value, as the identity of the Founder is unknown. All that is known is that the Founder appeared as a total stranger sometime in 1960, in a small town outside of Denver, and for just one day preached a doctrine of non-violence. This alarmed the local authorities, who were fearful of anybody 'strange.' (When Dick wrote this story, America was caught-up in the 'communists-are-everywhere' hysterical frenzy resulting from the political machinations of Wisconsin Senator Joseph McCarthy.[14]) So, they arrested and jailed him, murdered him while claiming he had hung himself, and then buried his body.

The crowd of people who heard him speak just before his arrest had been swept-up by his words and when, a few months later, some claimed to have seen him once again alive, rumors quickly spread that he had defeated death, and that he was divine. The Movement grew, resulting in the creation of the First Church, a continuing threat to the ruling authority that does not value non-violence. This threat has become so serious, that it has been decided to stop it before it can begin, by killing the Founder *before* he speaks. While the identity of the Founder is unknown, First Church followers eventually had recovered his skeleton and preserved the bones as sacred relics. These have been stolen by the ruling authority and, in particular, Conger is given the Founder's skull—with distinctive front teeth—to use for identifying the Founder. Conger agrees to the deal, although his employers admit "There is some philosophical doubt as to whether one can alter the past. This should answer the question once and for all."

So, back Conger goes to April 5, 1961. He intentionally picks a date *after* the Founder is known to have made his sermon, to allow him to search back issues of the local newspaper for a report of the arrest and death of a stranger. He finds what he is looking for, and so pin-points both the where and the when of the Founder's arrest (December 2, 1960, just 4 months earlier). As Conger heads back to his hidden time machine to make that 4 month jump

---

[14] When Conger arrives in the past he has a beard. When one of the local townsfolk sees that, she says "Why does he have a beard? No one else wears a beard. Must be something wrong with him. Wait. Didn't that—what was his name? The Red—that old one. Didn't he have a beard? Marx. He had a beard."

further back in time, he is surprised to observe a strange reaction in people who see him. Afraid that there has been some mistake in either his manner of speaking or dress, a potentially fatal mistake that might tag him as anachronistic, he hurries away and quickly makes an escape with his second jump back through time.

As he waits in hiding with a gun for the Founder to appear, Conger idly examines the skull. Suddenly, struck with an odd thought by what he sees, he stands before a mirror. Holding the skull beside his head, he bares his teeth. They match the skull's. There is no need to wait any longer; *he* is the Founder, and Conger realizes what the disturbance was all about just before he made his second time jump. To the people 4 months hence, it will appear that he has come back from the dead (as thought those who saw Jesus after the Crucifixion). With admirable resignation, he goes out to complete the events that history records, including his own "death foreordained." His words to the gathering crowd are few but powerful:

> "I have an odd paradox for you," he said. "Those who take lives will lose their own. Those who kill, will die. But he who gives his own life will live again!"

The time loop Dick describes is a self-consistent one (don't overlook the irony in Conger's last sentence), and while Conger has indeed played a central role in the past he has not changed the past. As he thinks to himself as the police come forward to take him away, "It was a good little paradox he had coined. They would puzzle over it, remember it."

In a certain sense, the greatest use of time travel to affect the past occurs in the 1941 story "The Seesaw" by A. E. van Vogt (1912–2000). There we learn the past itself was created, indeed all of time was created, not by God but rather by an accidental time traveler. Inadvertently getting caught-up in political intrigue between adversaries in the far-future (the overly complicated way this is explained is best avoided here!), a man from 1941 finds himself swinging through time from past to future to past to . . ., with every swing more extreme than the last.[15] With each swing he becomes "charged with trillions of trillions of time-energy units" and eventually "the stupendous temporal energy" will cause him to explode, deep in the past. That explosion will be what we call the Big Bang beginning of the universe. (In an introduction to this story, Isaac Asimov observes that van Vogt actually says that the deep-past explosion will

---

[15] The idea of a time traveler oscillating in time appears in the 1969 best-seller *Slaughterhouse-Five* by Kurt Vonnegut (1922–2007), describing the adventures of Billy Pilgrim. Vonnegut's work had a strong SF flavor to it, and it is almost certainly the case that he had read van Vogt's story years earlier.

create *the planets*, but that if he were writing the story today Asimov would make it the Big Bang.)

If one is willing to entertain the possibility of changing the past—an idea I personally think without logical support *in a universe with a single time line,*[16] an objection that I think should *not* kill a good story idea—then the only limit is the author's imagination. As an illustration of how Dick's imagination was up to this challenge, just 2 years after writing the 'unchangeable past' story "The Skull," he wrote "Jon's World" which takes exactly the opposite position. This tale opens with a description of a world that has been devastated by planet-wide war. Earth is covered by endless ruins rising out of vast expanses of ash: the surviving cities that are being rebuilt resemble "occasional toadstools." The destruction was the result of what was actually *two* wars: the first was of men against men, and the second was men against the intelligent robots that one side had initially developed as a weapon for the first war and which then turned on their makers.

Ironically, to help rebuild the post-wars world the aid of such robots is needed but, unfortunately, the technology behind the robot's artificial brains was lost when the last one was destroyed in the second war. The work of the inventor of the artificial brain has proven impossible to duplicate, and so is born Project Clock in which time travelers will journey back to before the start of the wars and simply steal the inventor's research notes. While the reader is being informed of all this, we also learn that one of the time travelers has a son, Jon, who periodically suffers from visions of an idyllic world, one free of the evils of war, and that these visions are growing ever more vivid as the day of departure to the past nears.

Then the trip begins, and at first it appears to be successful. The time travelers locate the inventor at a secret, heavily guarded government facility, and steal his research notes. As the travelers make their escape back to their time machine, however, the inventor joins the soldiers giving pursuit and is killed in an exchange of gunfire. The inventor's death is not part of recorded history, and so they suspect they have changed the past and thus have changed the future. They fear the world they left from may be quite different from the world to which they will return.

And they are right—the world they return to is the wonderful world that Jon saw in his visions, and the time travelers are stunned. Jon's father, in particular, realizes that "his world no longer existed . . . Jon. His son. Snuffed

---

[16] General relativity gives physics the theoretical possibility of time travel along a single time line, but the introduction of quantum mechanics gives rise to the possibility of infinity of time lines (the so-called "many-worlds" view of reality). I'll discuss the theological implications of multiple time lines at the end of this chapter.

out. He would never see him again ... everything he had known had winked out of existence." His colleague, however, sees a happier theological interpretation of what they have done, and he argues that what Jon saw might explain "the mystical visions of medieval saints. Perhaps they were of other futures, other time flows." "Jon's World" is essentially a wish-fulfillment, feel-good fantasy (and not a very subtle one, at that), while "The Skull" is a *logical* SF story (and while also a feel-good tale it is far more subtle).

Changing the past is a concept that has captured the attention of many writers, and it can be traced back, in theological fiction, to before 1900. I discussed Hale's "Hands Off" in Chap. 2 and, in another early example of 'experimenting' with the past, Mark Twain had Satan give a good lecture on it in a preliminary draft (written before the turn of the twentieth century) of his last novel, *No. 44, The Mysterious Stranger*: "If at any time—say in boyhood— Columbus had skipped the triflingest little link in the chain of acts projected and made inevitable by his first childish act, it would have changed his whole subsequent life, and he would have become a priest and died obscure in an Italian village, and America would not have been discovered for two centuries afterward. I know this. To skip any one of the billion acts in Columbus' chain would have wholly changed his life. I have examined his billions of possible careers, and in only one of them occurs the discovery of America."

A swashbuckling 'change the past' story is the novel by John Brunner that I mentioned earlier, *Times Without Number*. The novel opens in 1988, in a world far different from 'our' 1988. It's the world that would have resulted if, 400 years earlier, the Spanish Armada had *defeated* the English. It's a very strange world to 'our' eyes, as the novel imagines a social structure that is very much like that of *The Three Musketeers*; men wear velvet breeches and wear swords, women swoon, land travel is by horse-drawn coach, and long-distance communication is via semaphore telegraph. There are no cars, no radios, no space travel (which one character says would be a "miracle" if ever accomplished), but there has been time travel since 1892!

Time travel is tightly monitored in Brunner's novel by the Society of Time, under control of the Catholic Church in general, and specifically by Jesuit time cops.[17] The highly secret Reference Library of the Society contains thousands of theoretical analyses on all aspects of time travel, right down to the minutia

---

[17] The time police in SF, charged with thwarting those who would change history, are government agents who roam the corridors of time much like Marshall Dillon of *Gunsmoke* prowled the television streets of Dodge City, Kansas, and they represent bad physics. I agree with the Princeton philosopher David Lewis (1941–2001), who called their presence in a time travel tale "a boring invasion"—see his classic paper "The Paradoxes of Time Travel," *American Philosophical Quarterly*, April 1976, pp. 145–152, as well as Alasdair Richmond, "Time-Travel Fictions and Philosophy," *American Philosophical Quarterly*, October 2001, pp. 305–318.

of how a single, tiny, inadvertently created ink-blot on a Medieval manuscript (left by a careless time traveler) might change history. This library is under the control of the "the master-theoretician of the Society and the greatest living expert on the nature of time and the philosophical implications of travelling through it."

The central rule of the Society is "observation without interference," and that is maintained by forbidding time travel to all but the time cops of the Society. An exception is made, however, for every newly elected pope, who is allowed a trip back to the ministry of Jesus to assure the new Bishop of Rome that Jesus was not a mere historical figment. (Faith alone, apparently, being not *quite* enough even for the Holy Father!)

The novel has a surprisingly down-beat ending, but one that obeys Niven's Law. The world that opens the novel, because of careless changes made in the past, is with a single exception eliminated and replaced with 'our' world in which the Spanish Armada lost. The single exception is that one of the Jesuit time cops becomes trapped in 'our' 1988 and is shocked at how time travel has literally obliterated all that he knew. Even though he could become wealthy and famous in 'our' world with his knowledge of time travel ("he could describe the principle of time apparatus; given a ton of iron and half a ton of silver he could build [a time machine] with his own hands in a week"), he vows he will remain silent. In true religious penitence, he accepts his fate as "the most isolated of all the outcasts the human race had ever known."

A more recent, funny theological tale of this sort is "Pebble in Time" from the August 1970 issue of *Fantasy and Science Fiction Magazine*. In it we read of how an Elder of the Church of Latter-Day Saints invents a time machine so he can go back to 1847 to watch Brigham Young declare "*This is the place!*" at what would become Salt Lake City. Inadvertently interfering with the past, however, the traveler is shocked to instead hear "*This is not the place! Onward!*" and to then watch Young continue on to San Francisco. As the home of the Mormon Church in this altered history, San Francisco becomes associated with the initials L.D.S.—and so we see the story is simply a play on the initials our history associates with the permuted initials L.S.D. (the infamous mood-changing hallucinogen lysergic acid diethylamide of San Francisco's drug culture).

Far more serious in tone is the 1982 story "Angel of the Sixth Circle" by Gregg Keizer. Here we find a time traveling assassin named DeVries who serves a future, science-based new Church, one in murderous (literally) theological conflict with Catholicism. Sent on targeted missions of death into the past by the leader of the new Church, the Most Reverend, this religious hit-man strikes down all those Catholics whose elimination from history, it is calculated, will promote the interests of the new Church in the future. This killer isn't the only one who can time travel, however. First, it is a crime to be a time tamperer, and so he must constantly be on the look-out for operatives of 'the Sanction' (that is, the *time police*). And then there are the time assassins of the Catholic Church, too, staffed (as you might suspect) by the science-oriented Jesuits.

The story centers on the current assignment of DeVries, the *prevention* of the killing of Tomás de Torquemada (1420–1498), the Dominican friar who was the first Inquisitor-General of the infamous Spanish Inquisition. He is sent on his way, backward through time, with these words from the Most Reverend: "Do not let that butcher die, DeVries." It may seem an odd task for a man who normally kills Catholic priests in the past, but the reasoning is that the Jesuits have come to believe that the death of Torquemada would lessen the impact of the coming Reformation, that his elimination would result in one less excess of the Catholic Church for people to rebel against. In addition, all the Catholics that the Inquisition would burn at the stake in the present reality would, in a new reality, live to produce more Catholics. With Torquemada's death the authority of the Catholic Church in general, and of the Pope in particular, would not be questioned to the degree it was in the present reality. Thus, the Catholic Church puts out a 'contract-hit' on its own man.

The new Church, of course, wants to defeat the Jesuit plan and to have the hated Torquemada survive to continue the burning alive of heretics,[18] and so it is DeVries' task to kill the Jesuit assassin before that assassin can kill Torquemada. In this he fails—he does kill the Jesuit, but only after the assassin has succeeded in *his* task by strangling Torquemada. And that is when he accepts the truth of his own condition—he doesn't care that he failed, or that Torquemada's 'premature' death alters nothing—he cares only for the fact that he likes to kill. As he thinks to himself, "My faith was no longer my religion, in my Most Reverend, not even in God. It was in myself, the knowledge that I could kill with impunity, wanted to kill, and that here [in the new Church] I had an outlet for that, a place where it was sanctioned. Indeed, revered." And most chilling of all, is when the Most Reverend tells him "Your soul means nothing to the [new Church]. But as long as you have a faith, you will continue to do God's bidding."

To kill.

We actually do not have to turn to SF to find theological interest in the possibility of affecting the past. Quite respectable, real-world theologians have

---

[18] This explains the irony of the story's title, as the sixth circle of Dante's *Inferno* is where heretics are punished. The assassin is no angel to heretics, but rather to Satan, for enabling Torquemada (or at least attempting to enable him) to send even more souls to suffer in Hell.

long believed in such a thing, with their concept of what is called the *retroactive petitionary prayer*. An 'ordinary' petitionary prayer, like the Lord's Prayer in *Matthew* 6 or *Luke* 11, asks for something in the present or the future, while a retroactive prayer asks for something in the past. Two examples of retroactive prayers are the surgical patient who prays, just before an exploratory operation, that a suspected tumor to be non-malignant, and the soldier's wife who prays that her husband wasn't among those killed in yesterday's battle. These prayers are for a happy outcome to an event that was decided *before* the prayer is made. One might accept the rationality of praying about the future ("Please, God, let me survive tomorrow's battle and I'll be good for the rest of my life"), but are prayers about the past even sensible?

C. S. Lewis answered that question as follows[19]:

> "When we are praying about the result, say, of a battle or a medical consultation, the thought will often cross our minds that (if only we knew it) the event is already decided one way or the other. I believe this to be no good reason for ceasing our prayers. The event certainly has been decided—in a sense it was decided 'before all worlds.' But one of the things taken into account in deciding it, and therefore one of the things that really causes it to happen, may be this very prayer that we are now offering. Thus, shocking as it may sound, I conclude that we can at noon become part causes of an event occurring at ten A.M. (Some scientists would find this easier than popular thought does.)"

With those words it is clear Lewis believed that the present could indeed affect (but not change) the past. His last sentence in the quote shows that he realized that he wasn't alone in that view, and that (for once) his theology and modern science shared the same position on a technical issue. Lewis never explicitly mentions the block universe, but it seems equally clear that he believed in the idea of God being able to see all of reality at once, and that God knew of the petitionary prayer before it was made. Or, to put it in even stronger terms, that God is not a temporal being but rather is 'eternal' and so knows all of time 'at once.' That is, God knows of the prayer and the event being prayed about 'at the same time.'

Lewis does make it clear that he believed it to be a sin to pray for something *known* not to have occurred. As he wrote (in *Miracles*),

> "If we can reasonably pray for an event which must in fact have happened or failed to happen several hours ago, why can we not pray for an event which we know not to have happened? e.g. pray for the safety of someone who, as we

---

[19] See the appendix titled "On Special Providences" in Lewis' book *Miracles*, Macmillan 1978.

know, was killed yesterday? What makes the difference is precisely our knowledge. The known event states God's will. It is *psychologically impossible* [my emphasis, and what Lewis meant by this escapes me] to pray for what we know to be unobtainable; and if it were possible the prayer would sin against the duty of submission to God's known will."[20]

What's *logically* wrong with this cartoon? (The answer is at the end of this chapter.)

Cornered
by Mike Baldwin

12-17 © 2005 Mike Baldwin / Dist. by Universal Press Syndicate www.cornered.com
cornered@comic.com

There it was: the same piece of cake he ate yesterday. His time-machine really worked. Think of the possibilities. He could have his cake and eat it too.

The struggle between a fixed and a malleable past is beautifully illustrated in Robert Frost's famous 1916 poem "The Road Not Taken." It opens with "Two roads diverged in a yellow wood/And sorry I could not travel both." Then later come the lines "And both that morning equally lay/In leaves no step had trodden black./Oh, I kept the first for another day!" Could these words be interpreted to mean one could later return and "do things" differently? The ending of the poem, however, makes it clear (I think) that Frost was consciously thinking of the crucial (unchangeable) nature of

---

[20] For more on this issue, see T. J. Mawson, "Praying for Known Outcomes," *Religious Studies*, March 2007, pp. 71–87. What I think Lewis meant by *psychologically impossible* is that such a prayer could not be a *sincere* prayer.

decisions: "Two roads diverged in a wood, and I—/I took the one less traveled by,/And that has made all the difference."

7.3 Jesus and Time Travelers

With time travel, all sorts of interesting adventure possibilities open-up for the SF writer. The past, itself, would be the ultimate tourist attraction, with not just places but also *dead people* to visit, people who become alive 'once more.' People like Lincoln, Hitler, and of course (especially for us, in this book), Jesus. With Lincoln, the goal is usually to either save him from assassination (doomed to failure if the past is unchangeable[21]), or for purely scholarly reasons. For example, in the 1958 novel *The Lincoln Hunters* by Wilson Tucker (1914–2006) we read of a business called Time Researchers. T-R recovers lost historical artifacts; specifically, an original sound recording is made of one of Lincoln's speeches.

When Hitler appears in a time travel story, however, the goal is far less benevolent (surely that is no surprise); the goal is almost certainly to kill him before he rises to power. (Writers have been attempting to get rid of Hitler, even when he was still alive, ever since Geoffrey Household's 1939 novel *Rogue Male;* with time travel, even being dead *now* isn't sufficient to spare him from would-be assassins.) In the short story "The Plot to Save Hitler" (*Analog Science Fiction Magazine,* September 1993), for example, a time traveler journeys back to 1904 to kill the then 14-year-old Hitler. A quite interesting exception to this is the 1994 story "Inspiration" by Ben Bova (born 1932), and I'll return to it in the next section.

I specifically mention Hitler because, while there doesn't seem to be anything at all religious about the monster he became as an adult (other than as an agent of evil), there is clearly a *moral* issue that must be addressed. If a time traveler confronts the boy Hitler, wouldn't it be a sin to kill him *before* he has committed any crime? Some might argue that it is okay to kill him because of the time traveler's knowledge of what Hitler *will* do in his future— but that's assuming a role very nearly God-like. And, after all, even with *His* omniscience, God did not prevent Hitler's crimes against humanity.

Such stories of Lincoln and Hitler are interesting, but they cannot compare (in my opinion) with the sheer mystery that automatically comes with a time trip back to Jesus. For example, in "The Rescuer," a 1962 story by Arthur Porges (1915–2006), we read of a man, in the year 2015, who takes a rifle and

---

[21] See, for example, Robert Silverberg's 1957 story "The Assassin," in which an attempt to save Lincoln from Booth is foiled by bodyguards who mistake the time traveler's portable time machine for a bomb, and so destroy it and haul the would-be savior off to jail.

5,000 rounds of explosive bullets back in time to Golgotha. His intention—to be history's first Rambo by picking off any Roman soldier who gets within a hundred yards of Jesus! As outrageous as this concept is (but who among readers wouldn't admit to at least a momentary thrill at the idea, and perhaps a secret willingness to do it themselves, if they could), it isn't the story's peak. That comes when we are reminded that it was Jesus' *desire* to die on the Cross, that he *had* to die for our sins; to prevent that from happening would subvert Jesus and change all of history for the last 2,000 years. What, then, should the time traveler's colleagues do when they understand his intent? Should they somehow stop him?

Those questions are perhaps not so easy to answer. Here's why. The instant after the armed time traveler leaves for the past *he has been in the past for 2,000 years*. Indeed, has been in the past even *before* the time machine was built! Whatever he did there (then) has been done for 2,000 years. So, just *how* do his colleagues stop him? Well, you might suggest, how about they go back to the day *before* the Crucifixion and, when Rambo appears the next day, that's when they stop him. Okay, but again, even before the colleagues begin their chase *they* will have been in the past for 2,000 years, too. But wait, if they 'remember' the Crucifixion, then of course the past *didn't* change—and so then why bother going back? But wait—maybe it's a new memory that, of course, they just *think* is what they have always remembered. And so they *should* go back. Or, maybe . . . . Oh, my, yes, you *can* become deranged thinking about such things![22]

Not all the time travelers in SF seek Jesus with love in their hearts. In the 1922 "Un Billiant Sujet" by the French writer Jacques Rigaut (1898–1929), for example, a time traveler commits many disturbed acts in the past, one of which is the murder of the infant Jesus with an injection of potassium cyanide. David Gerrold (born 1944) did something similar in his 1973 novel *The Man Who Folded Himself*, with a time traveler who experiments with changing the past. In the words of that character, "Once I created a world where Jesus Christ never existed. He went out into the desert to fast and never came back. The twentieth century I returned to was—different. Alien."

Somewhat paradoxically, perhaps, there are religious groups in SF who don't want *anybody* to seek Jesus in the past. In the 1978 novel *Mastodonia* by Clifford Simak, for example, the owner of a 'time travel portal' to the past is visited by representatives of such a group, who explains this curious position:

---

[22] The situation as I've described it is called a *bilking paradox*. See *Time Machines* (note 3), pp. 196–197 and 332–336, for how physicists respond to this puzzle. In Porges' story, the would-be savior is discovered while the time machine is still powering-up, and the time machine is destroyed before the journey is actually made. Would *you* destroy the machine?

"There always has been a question of the historicity of Jesus. Nothing is known about Him. There are only one or two literary mentions of Him and these may be later interpolations. We don't know the date or the place of His birth. It is generally accepted He was born in Bethlehem, but even on this, there is some question. The same situation holds true in every other phase of His life. Some students have even questioned the existence of such a man. But through the centuries, the myths that have been brought forward regarding Him have been accepted, have become the soul, the structure, the texture of the Christian faith. We want it left that way. . . . What do you think would happen if it were found that Jesus was not born in Bethlehem? What would that do to the Christmas story? What if no evidence were found of the Magi?"

Oddly, the possibility of obtaining knowledge of the truth means nothing to them, as the same character goes on to say "[We] are [not] men of little faith. The truth is that our faith is so all-encompassing that we can and do accept Christianity even knowing that little is known of Our Lord and that little may be wrong." Religious ignorance, in this novel, *is* bliss.

There are time travelers in SF who go back to the Crucifixion just as modern vacationers go to Disney World—as one of the tourists Hawking wondered about earlier in this chapter. A famous story of this type is the 1975 tale "Let's Go to Golgotha!" by Garry Kilworth (born 1941). After commercial time travel has been invented the story imagines that, among other famous historical events, you can book a trip to the Crucifixion (the Coronation of Elizabeth the First, and the Sacking of Carthage, are quite popular, too). Taking the whole family to see the death of Jesus has become not much different from catching the early afternoon matinee at the local movie theater. Before leaving for the past, all tourists on the tour are warned that they must avoid doing anything that will change recorded history. In particular, when the crowd is asked whether Jesus or Barabbas should be spared as the single amnesty allowed during the Feast of the Passover, all must shout "Barabbas!" That is shocking in itself, but not nearly as much as occurs when a tourist realizes that the entire crowd around him, with all condemning Jesus to death, is *nothing but* other tourists from the future, and that in fact there are no inhabitants of 33 A.D. present (other than the condemned Jesus and His Roman soldier executioners). The locals are in their houses, quietly praying, while it is only the time tourists who are dancing in the streets.

A similar story, written before Kilworth's because I recall reading it in the mid-1950s (but I cannot recall either the author or the title), is also of a tourist in disguise at Golgotha for the Crucifixion. He has a camera, but of course he must hide it beneath his robe to avoid attracting attention. All goes well until he notices clicking noises coming from those standing near him. It is then he realizes the entire crowd is *nothing but* time travelers, from all through the ages, all with hidden cameras beneath *their* robes! In Kilworth's tale the tourists are active (if unhappy) participants in the killing of Jesus, but it strikes me that the *indifference* inherent in a tourist taking a camera back to the Crucifixion is—somehow—just as awful.

A different sort of Crucifixion observer from the future is the professor in the 1954 story "The Traveller" by Richard Matheson[23] (1926–2013). A skeptic (he thinks the Crucifixion is 'a fallacy of the ages') as he begins his chrono-transposition journey 2,000 years into the past, he arrives on Golgotha at 9 o'clock in the morning of the day of execution, a half mile outside the walls of Jerusalem. His initial reactions are dreary ones. The place, looking "something akin to an unkempt city lot," is littered with animal excrement and garbage in which dogs are foraging, and it stinks. The Cross is nothing at all grand (as pictured in churches), being simply an upright stake and a cross beam crudely lashed and nailed on top. The feet of Jesus will be just inches from the ground but, of course, that will do the job just as well as if he were suspended (as usually imagined) many feet high above an awe-struck crowd. And speaking of a crowd, there isn't one—the place is deserted except for a few bored Roman soldiers.

There are none of the portents, signs, or miracles mentioned in the Bible, no seamless robe, and the professor proclaims such tales to all be nothing but "Biblical drivel." The professor does spot a man who appears to be the one said to have helped Jesus when he fell while staggering towards the execution site—Simon of Cyrene—but the professor's skepticism nevertheless seems justified when he doesn't see either John or Mary of Magdalene. He does see Jesus, however, and the first impression is a strong one—Jesus is tall, very thin, dirty but healthy, with a handsome face—but there is nothing at first to mark him as anything special.

And yet—there *is* the sign the Bible mentions, in Greek, Hebrew, and Latin, declaring Jesus to be King of the Jews. The professor watches in shock as Jesus and his two thief companions are hung on their crosses. When Jesus utters his first word since appearing, he speaks of God, and despite himself the professor feels a powerful urge to help. He still thinks that the Bible stories of the Crucifixion are "all rot," but seeing a starved man stripped naked, lashed hands and feet to hang from a cross with nails driven through his palms, his face white with pain—and yet with sorrowful eyes filled with gentleness—causes the professor to attempt to stop the horror.

The time machine operators 'in the present,' who are in voice communication with the professor (just how *that* is done is called 'willing suspension of

---

[23] Best known in SF for his 1975 romantic time travel novel *Bid Time Return*, which was made into the 1980 film *Somewhere in Time*. Both the novel and the film make good use of loops in time.

disbelief!), can't allow such interference with the past, of course, and quickly bring the professor back. Yanking him back rapidly, however, results in a spacetime disturbance so violent that Golgotha experiences an earthquake, just as reported in the Bible. So, maybe, the stories aren't *all* rot. Best of all, for those who enjoy 'happy endings,' the professor's experiences result in him finding God and he is no longer a skeptic.

Finally, this section on time traveling back to Jesus would be incomplete without mention of what I think the most provocative (some would say outrageous) religious time travel story of all, the 1969 "Behold the Man" by the English writer Michael Moorcock (born 1939). Catholics will almost surely be outraged at its premise, and yet it is a masterpiece that combines careful historic research with a deep understanding of the psychology of an emotionally disturbed time traveler. The novella opens with time traveler Karl Glogauer arriving in the Biblical past, the very years of the young Jesus. His target is A.D. 29, the year before the Crucifixion (the actual date of the Crucifixion is still debated by historians). His time machine is a sphere filled with a cushion of milky fluid which spills through a crack as it makes a 'hard landing.' Karl emerges from the machine just like a new-born baby emerging from the womb into a new world.

There is then the first of many flashbacks,[24] to when Karl was nine, playing a schoolyard game with other children, with Karl as Jesus tied to a fence, and so the reader gets a first glimpse of his obsession with Jesus. It soon becomes clear that Karl is an unusually sensitive, highly emotional, sexually frustrated boy who has suicidal inclinations.

After Karl exits the time machine he faints from his painful injuries; when he wakes he finds that he has been discovered, patched-up, and carried to a building with a straw floor upon which he has been placed. Since Karl, in preparation for his time trip, spent 6 months in the British Museum studying the ancient Aramaic language of Jesus' day, he is able to communicate with his rescuers. He learns that Tiberius sits as Emperor in Rome, in whose reign Jesus was executed, and so he knows that he at least has reached the right time period. When asked where he is from, he leads his questioner to assume Egypt. Further questioning leads to being asked his name, and he replies with caution, knowing that 'Karl' would seem outlandish to Biblical people; he instead gives his father's name, 'Emmanuel,' realizing only afterwards it is Hebrew for "God with us." His questioner then reveals *his* name to be John the Baptist. That is good news for Karl, because the New Testament says the Baptist was killed— decapitated by Herod—*before* the Crucifixion. So, Karl is at the right time, but

---

[24] Actually it might be more accurate to call them flash-*forwards*, since Karl is in the past when we read these passages describing his life *before* he makes the time journey.

he is also puzzled because John says he does not know of anybody called 'Jesus the Nazarene.'

But what really amazes Karl is that the Baptist believes *Karl* is the long-prophesized messenger (messiah) from God! Karl's incredible appearance in the time machine was seen by a number of people; as John tells him, "Are you not a magus [an ancient term for a man with special powers—each of the Three Biblical Wise men in the Gospel of Matthew was a magus, and collectively they were the Maji], coming in that chariot [the time machine] from nowhere? My men saw you! They saw the shining thing take shape in the air, crack and let you enter out of it. Is that not magical? . . . The prophet said that a magus would come from Egypt and be called Emmanuel. So it is written in the Book of Micah!" (The reference is to an eighth century B.C. prophet whose words appear in the Old Testament.)

Eventually Karl breaks away from John and, alone, strikes out in search of the town of Nazareth, and of Jesus. He finds both, but is shocked when he discovers that Jesus, the son of the carpenter Joseph and his wife Mary (just as the Bible says), is a misshapen, congenital imbecilic simpleton idiot who wets the floor. As Karl stares in horror at this very non-Biblical Jesus, Mary and Joseph engage in a heated argument over the 'story' Mary told her parents about the origin of Jesus. It involved a wild tale about being 'taken by an angel,' a story that nobody really believes but finds more convenient to accept than would be the truth.[25]

Karl is taken in by the rabbis of a Jewish synagogue who ask where he is from, and he replies "The world to come and the world that is," meaning the future—but the rabbis think he means 'the next world' after the coming of the Lord. They believe Karl to be a holy man and they clean him up, nurse him back to health, and give him the run of the synagogue. During conversations, he occasional replies in English without thinking and the rabbis are tremendously impressed by the incomprehensibleness of his words!

Karl, in turn, is astonished at the generally wide-spread suffering from psychosomatic illnesses (such as hysterical blindness), and finds that his reputation as a holy man allows him to perform 'miracle cures' simply with the laying on of hands. Some of his utterances are interpreted as profound predictions, such as the arrest soon after of John the Baptist by Herod. Karl is happy and excited by the respect such acts give him, an acceptance by others

---

[25] At this point you might begin to suspect that Moorcock thinks little of the 'divine' origin of Jesus. One has to be careful not to confuse an author's actual beliefs with what his fictional characters say, but in this case you'd be right. Moorcock once wrote, in an afterword to an abbreviated version of "Behold the Man," "I meant no offense to religious people, and in fact had no overtly religious acquaintances. Such people did not exist in my circle, and it was not until I came to America that I realized religion was not dead—for me, a type of time travel, like going back to the Dark Ages."

that he never enjoyed in his earlier life. Since it was known that he was somehow associated with Joseph and Mary—something to do with somebody named Jesus—the name Jesus becomes associated with Karl, and many begin to proclaim him to be the prophesized messiah.

Karl knows what is happening, but denies to himself that he is in any way 'changing history.' Rather, "he was merely giving history more substance." (Karl, of course, is *changing* nothing, but is simply an 'actor' of sorts in a time loop.) Remembering the Biblical tales, Karl does his best to replicate them. When he is asked by a follower of John the Baptist, for example, to plead with Herod to spare John's life, Karl/Jesus replies "He must not be helped. He must die." When John is eventually beheaded, Karl's influence only grows. Eventually, to complete his self-appointed mission of bringing the Biblical Jesus to life, Karl arranges for Judas to 'betray' him to Pontius Pilate (thus, in Moorcock's story, Judas is not a traitor for silver but, instead, is actually doing the bidding of Karl/Jesus and so has been given a bum rap in the Bible for 2,000 years[26]).

The novella ends (as we know it must) with Karl/Jesus being arrested, tried, and executed on the Cross. As he dies he knows the truth, that he is no Son of God but rather is simply a neurotic time traveler from the future, and so the last words on his lips, as he suffers horribly, are "It's a lie—it's a lie—it's a lie . . . " This brutal story ends with Karl's corpse rotting in the dissection room of a doctor who stole it from its tomb (thus fulfilling the final Biblical story of Jesus' life on Earth) because he thought it might have some special property. It didn't, and it was soon destroyed, secretly buried, and its worldly fate forgotten.

It is interesting to compare "Behold the Man" with Dick's "The Skull." Both deal with a time traveler whose trip into the past is part of a self-consistent causal time loop, with each becoming the very individual they are seeking. On a deeper level, however, the two tales couldn't be more different. Dick's is actually uplifting, with the First Church founded by the time traveler seeking to defeat the ruthless political authorities that sent him on his path into the past. Moorcock's story is a sad one for all (and devastating for those who embrace the divinity of Jesus), with its premise that the traditional beliefs are, in Karl's dying words, 'nothing but lies.'

The idea of using time travel to meet Jesus continues to appear in SF, with a recent example being the story "Salvation" in the December 2007 issue of *Analog*. In it a scientist appeals to the Universal Church of the Divine

---

[26] This is not an original idea with Moorcock; it can be found, for example, in the 1944 essay "Three Versions of Judas" by the Argentine writer Jorge Luis Borges (1899–1986). It is included in his famous 1944 anthology *Ficciones*.

Revelation for financial support so he can build a time machine. The cynical irony is that the scientist is a religious skeptic, but he knows where the money is. This story does not follow Moorcock's logical development of the time traveler's actions in the past being part of a self-consistent causal time loop. Rather, it allows the scientist to change the past, even though it is stated that there is just *one* time dimension. So, after a heart-to-heart chat with Jesus in which the scientist encourages a receptive Son of God to lighten-up on the mysticism and to include more science in his ministry, the time traveler returns to what the author clearly implies is a 'better' present.

7.4 Quantum Mechanics and God

Back in Chap. 2, in the discussion of Hale's story "Hands Off," you'll recall I mentioned in passing the many-worlds view of quantum mechanics, and Leinster's 1934 story "Sidewise in Time." That story is a "parallel universe" tale that uses general relativity's spacetime warping property to create new universes separate and distinct from 'ours.' The introduction of quantum mechanics, *in a world with time travel*, does something entirely different. If a time traveler in such a world journeys into the past and introduces a change (indeed, the journey itself may be the change) then reality splits or *forks* into two versions. One forking path in spacetime is the result of the change, and the other forking path is the original reality without the change.

This view has the immediate result of eliminating paradoxes: one *can*, in this view, kill your grandfather, sort of. On one path the killing fails and you will be born in the subsequent ('original') reality, and on the other path you succeed and so you won't be born in the subsequent ('new') reality. The usual paradox doesn't appear because the murderous you in the 'new' reality comes from the *other* reality! SF has enthusiastically adopted quantum mechanics and its connection with time travel. An example of this view of time is Ben Bova's "Inspiration," that I mentioned briefly in the previous section. In it a time traveler from a horrific future surreptitiously arranges a meeting in the summer of 1896 (the year after the publication of *The Time Machine*) between himself, H. G. Wells, and the then 17 year-old Albert Einstein. The three have just attended a lecture by the famous Professor William Thomson (Lord Kelvin) and heard him make his equally famous (and false) claim that everything was now known in physics, a claim that has greatly discouraged Albert from pursuing physics. The meeting (Lord Kelvin soon joins them) takes place in an Austrian café in Linz; the waitress is the mother of Adolf Hitler and she exhibits great hostility towards the Jewish Einstein. Hitler, a young boy of 6, works there, too.

We learn of the time traveler's origin after Wells complains about Linz (narrow streets, bad food, etc.) and the traveler thinks "I, of course, knew several versions of Linz even less pleasing, including one in which the city was nothing more than charred radioactive rubble and the Danube so contaminated that it glowed at night all the way down to the Black Sea." And we learn the reason for his trip into the past when we read "There was only one timeline in which Albert lived long enough to make an effect on the world. There were dozens where he languished in obscurity or was gassed in one of the death camps," and so the time traveler attempts to reignite Albert's interest in physics by giving him a copy of *The Time Machine*.[27] The time traveler knows who the young Hitler is, too, and what he'll become, and simply thinks to himself that his efforts must be to "save as much of the human race" as he can but, as for Hitler, "it was already too late to save him."

The first such 'branching timelines' tale was "The Branches of Time" by David R. Daniels (1915–1936) that appeared in the August 1935 issue of *Wonder Stories*. This pioneering story contained the important observation that while forking time tracks may allow for changing the past for the better (something that, logically, can't be done for better *or* worse in a world with just one time track), in the end such a change may be futile. As Daniels' time traveler puts it, "I did have an idea to . . . go back to make past ages more livable. Terrible things have happened in history, you know. But it isn't any use. Think, for instance, of the martyrs and the things they suffered. I could back and save them those wrongs. And yet all the time . . . they would still have known their unhappiness and their agony, *because in this world-line those things have happened* [my emphasis]. At the end, it's all unchangeable; it merely unrolls before us." As Bova's time traveler laments, "Time branches endlessly and only a few, a precious handful of those branches manage to avoid utter disaster."

The above words[28] from Daniels' story have an interesting theological implication, one first raised by a philosopher.[29] Arguing that God cannot

---

[27] In the interest of telling a good tale, Bova has taken a few historical liberties; for example, there is no historical record of Hitler's mother ever working as a café waitress, Hitler was 7 (not 6) in the summer of 1896, and the origin of Einstein's work on relativity is known from his own words to *not* have been Wells' novel. Bova would have an easy answer to these quibbles, however: his time traveler is simply on a timeline different from ours!

[28] The sensitive passage I quoted is particularly impressive when you realize Daniels was no more than 20 years old when he wrote it. He soon after committed suicide and a promising writing-life came to a tragic end.

[29] Quentin Smith, "A New Topology of Temporal and Atemporal Permanence," *Nous*, June 1989, pp. 307–330.

branch along multiple time tracks because God is unique, philosopher Smith concludes that God can exist only in exactly one of how ever many time tracks there may be. What if that special time track isn't ours? Then, philosopher Smith concludes, Nietzsche's famous nineteenth century metaphorical claim that "God is dead" (or at least He is on a different time track) might be literally true. He admits that this is "fanciful," but still . . . .

Quantum mechanics and time travel can even 'explain' the origin of the Universe! Here's how. In the previous chapter (see note 25) I mentioned Stanislaw Lem's talent as a writer of convincing scientific SFBS double-talk ("high-class faking," one reviewer more graciously called it[30]). Lem was a master at writing satiric SF using that skill, and nothing else that he wrote demonstrates it better than his 1981 "Project Genesis." It is a tale so witty that it charmed the demanding editors of the elegant *The New Yorker Magazine*, who gave the story its first English appearance (in the November 2, 1981 issue).

It all begins by flipping the Heisenberg Uncertainty Principle of quantum mechanics on its head—that is, Lem states the Principle backwards, in the hope that nine out of ten readers won't notice. As he writes, "Mesons, those elementary particles, sometimes violate the laws of conservation, but they do this so incredibly fast that they hardly violate them at all. What is forbidden by the laws of physics they do with lightning speed, as though nothing could be more natural, and then they immediately submit to those laws again. . . . If mesons can behave impossibly for a fraction of a second, a fraction so minuscule that a whole second would seem an eternity in comparison, then the Universe, given its dimensions, might behave in that forbidden way for a correspondingly longer period of time. For, say, 15 billion years . . . ."

So that's Lem's starting premise: the Universe is a long-duration chance *forbidden fluctuation*' out of nothing. (Of course it's just the other way round—a mass the size of the Universe would be able to violate physical laws for a time interval far *smaller* than even the so-called *'quantum of time,'* 10^{-43} s. Nonetheless, let's go along with Lem—after all, this is SF! And so, in a single, brilliant stroke of insight (or rather, one of high-class SFBS), Lem has done away with God as the Creator of It All.

But, of course, in the interest of a good story, there *is* a problem. Such a quantum fluctuation represents a 'debt' to Nature that must, eventually, be repaid. As Lem puts it, the Universe exists on credit, and so one day it must burst like a bubble and return to its original non-existence. To eliminate the debt, all that is needed is to provide a *reason* (not just chance) for the fluctuation, that is, a *cause*. Then, no violation of physical law will (or *did*,

---

[30] Michael Kandel, "Two Meditations on Stanislaw Lem," *Science-Fiction Studies*, November 1986, pp. 374-381.

because you'll remember, we are talking of a past event) occur. The astonishing result of this realization is the creation of Project Genesis, which is an heroic attempt to prevent the 'bursting of our bubble,' so to speak.

The idea behind Project Genesis is 'simply' "to place a single solitary atom in the void, and the Universe could grow from it as from a planted seed, now in a totally legitimate way, in accordance with the laws of physics and the principle of the conservation of matter and energy." To do the planting, the narrator tells us that "we took a huge university synchrophasetron and rebuilt it into a cannon aimed at the beginning of time. All its power, concentrated and focused in a single particle—the constructional quantum—was to be released . . . from the Chronocannon." By a proper design of this seed, a design 'explained' by even more outrageous SFBS, it is thought an added plus is that a new, better world might be created than is the one we inhabit.[31] But all that fell apart when goofs were made by three members of the Project Genesis team (and so here we apparently have the famous Biblical Trinity)!

Thus, Project Genesis sort of worked and the Universe is not the result of a 'fluctuational caprice of Nothingness," but it also sort of failed, too, in that the world we have is still the same flawed world we had before the mighty Chronocannon fired its shot heard 'round the Universe.' It seems, you see, that you can't change the past, even in a quantum world.

A very funny use of quantum mechanics in a religious context was made by Mary Doria Russell in her novel *Children of God* (discussed in the previous chapter). There we find a character explaining to another an analogy between Schrödinger's Cat and the existence of God. This famous (in physics) thought experiment posits the placing of a cat in a sealed box, along with both a plate of good food and a plate of poisoned food. The cat may eat from either plate, and so it is either alive or dead inside the box, but until you open the box and look, all you can say is that each possibility has some probability. You have to *look* to make one of the possibilities *the* reality. The explaining character then reveals where he is going with all this: "Now here's my idea about God. I think we're like the cat. I think that God is like the man outside the box. I think that if the cat believes in the man, the man is there. And if the cat is an atheist, there is no man." What delicious irony this is if, like me, you think this is only fair considering how the Church continues to deny an immortal soul to the cat. In response, the cat simply does away with God!

Of course, it could possibly be far worse . . . .

---

[31] For example, one improvement was to be to "shorten interstellar distances, which would facilitate space travel and thus bring together and unify sentient races."

Saint Felix instead of Saint Peter?

To end this chapter (and to get back to time travel), let me observe that the concept of time travel is one that exists very nearly (but not quite as totally as does Lem's wonderful Chronocannon) in fantasy. There is *just barely* enough of scientific respectability to the known physics of time travel to keep it alive in SF. It all reminds me of a story told of the philosopher Sir Karl Popper (1902–1994), a wonderful story[32] about his apprenticeship in 1920s Vienna to a master cabinetmaker. After winning the old man's confidence, the student learned his mentor's great secret: for years the master had been looking for the solution to achieving perpetual motion. He knew the negative judgment that physicists had of such a machine, but he nevertheless refused to give-up his dream: "They say you can't make it, but once it's been done they'll talk differently." Might we one day say the same about time travel?

The theoretical support for time travel is just a *bit* stronger than is the total rejection by physics of perpetual motion so, maybe one day, *just maybe*, the first time traveler will propose a toast such as the one offered in "Time's Arrow" (the story I mentioned in the opening section) when the inventor of the first time machine and his no longer skeptical friend successfully arrive in the Civil War past:

"To you, Mac," I said.

McHugh loosened his tie. "To the Creator," he said, "who has given us a Universe with such marvelous possibilities."

---

[32] From volume 1 of *The Philosophy of Karl Popper* (P. A, Schilpp, editor), The Library of Living Philosophers, Open Court 1974, p. 3.

The Time-Traveler-and-the-Cake Puzzle

Every instant of a time traveler's life is his 'personal present,' and so the word *yesterday* is ambiguous. What calendar day the word refers to depends on where (when?) he was, in his personal past. From the caption's words "his time-machine really worked" we can assume that this is his first time journey, and that the start of the trip was both his *and the rest of the world's* present. Thus, as he stands in front of the refrigerator, he has jumped one day in time. But did he go *forward* one day or *backward* one day? If he went forward then 'yesterday' (when he ate the cake), was just before he started the trip. It is then clear that the cake could *not* still be in the refrigerator after the jump and thus the cartoon wouldn't make any sense. So, he must have gone backward one day, and his 'yesterday' is actually what all non-time travelers would call 'tomorrow.' Now, as he stands before the refrigerator and thinks of eating the cake (there it is, moist, tasty and available and he *really, really wants* to eat it because it was *so yummy good* when he ate it in his yesterday), the fact is that he *can't* eat it. That's because *it has to be there tomorrow* to be eaten *then*, just before he starts his trip. He *remembers* eating it then. So, the big question is: what prevents him from eating the cake *now*? Well, who knows *what* stops him, but *something must* to keep the time loop consistent. The 'problem' is that this argument seems to deny free-will, and *that* is what bothers people. We seem to have a brutal choice to make—either we give-up free-will, or we give-up logic. That's a tough choice, but nobody said time travel wouldn't be a queer business! This same sort of argument can be used to understand the answer to the famous grandfather paradox.

Chapter 8
What If God Revealed Himself?

"What is the point of Your being there if we know You only in
Your silence?"
—thought of Jesus of Nazareth as the mob comes for him, in
Jack McDevitt's "Friends in High Places"
"Ah, Lord, if I doubt You, it is perhaps because You hide
Yourself so well."
—a priest-physicist in Jack McDevitt's The Hercules Text

8.1 Not So Serious Speculations

In his great novel *The Brothers Karamazov*, Dostoevsky has one of his
characters (Mitya) say[1]

> "It's God that's worrying me. That's the only thing that's worrying me. What if
> He doesn't exist? What if . . . it's an idea made up by men? Then if He doesn't
> exist, man is the chief of the earth, of the universe. Magnificent! Only how is he
> going to be good without God? That's the question. I always come back to that.
> For whom is man going to love then? To whom will he be thankful? To whom
> will he sing the hymn? . . . [some say] that one can love humanity without God.
> Well, only a sniveling idiot can maintain that."

(Secular humanists, who believe that humans are capable of being ethical and
moral without any need for religion or God, would of course disagree.)

Decades later, in his 1928 novel *Point Counter Point*, English writer Aldous
Huxley (1894–1963) has one character telephone his brother with what he
calls a "really remarkable discovery" of a mathematical proof of God's
existence. It's the sort of thing that teenage schoolboys argue over, or at least

---

[1] Chapter 4 ("A Hymn and a Secret") of Book XI (*Ivan*).

P.J. Nahin, *Holy Sci-Fi!*, Science and Fiction,
DOI 10.1007/978-1-4939-0618-5_8, © Springer Science+Business Media New York 2014

until they take Algebra II: "You know the formula: *m* over nought equals infinity, *m* being any positive number? Well, why not reduce the equation to a simpler form by multiplying both sides by nought? That is to say that a positive number is the product of zero and infinity. Doesn't that demonstrate the creation of the universe by an infinite power out of nothing? Doesn't it?"[2]

Using mathematical arguments to 'prove' the existence of God long predates Huxley, and the practice was quite common among Victorian English clergy with a mathematical bent.[3] In his posthumously published 1872 book *A Budget of Paradoxes*, the great English mathematician Augustus De Morgan (1806–1871) recalled "When a very young man, I was frequently exhorted to one or another view of religion by pastors and others who thought a mathematical argument would be irresistible." An interesting example of such arguments, again using infinity, tries to compare life on Earth with the promised eternal happiness experienced in Heaven: Define the *infinite* value of the eternal bliss of the afterlife as x, and y as the *finite* value of the maximum possible happiness in life; since x + y = x (infinity plus anything finite is infinity), then the specific value of y is 'irrelevant'! (This could, I suppose, serve as an argument for continually giving all your discretionary wealth, beyond the bare minimum needed to keeping yourself alive, to the Church.)

America was also home to such arguments, which attempted to entangle mathematics with religion. For example, at the end of his 1851 treatise *Quadrature of the Circle* (by the New Yorker John A. Parker, republished in 1874 by John Wiley & Son) we find the statement "What are numbers? Before creation began, numbers had no existence, except in the infinite eternal One." Parker's 'credentials' as a mathematician are made obvious when you realize that, long after it had been proven that pi is irrational (actually, it's beyond being 'merely' irrational and pi is in fact transcendental), he claimed it is exactly $\frac{20,612}{6,561}$.

The desire to find a means for 'proving' God's existence is a strong one, and an appeal to the relentless, unemotional logic of mathematics is understandably powerful. As one character in Russell's *Children of God* says, "Abraham invented God because he needed to impose meaning on a chaotic, primitive world. We preserve this invented god and insist he loves us because we fear a large and indifferent universe." And earlier in the same work, yet another character—a Jesuit priest, no less—wonders (after a few alcoholic drinks) if "God is only the most powerful poetic idea we humans're capable of thinkin.' Maybe God has no reality outside our minds and exists only in the paradox of

---

[2] *Point Counter Point*, Harper & Brothers 1928, p. 135.a.
[3] Daniel J. Cohen, *Equations from God: pure mathematics and Victorian faith*, The Johns Hopkins University Press 2007.

Perfect Compassion and Perfect Justice." How seductive it is, then, to turn to the Queen of the Sciences, to mathematics, in attempting to show that God is not invented, but instead that He is 'built into' the very structure of the universe.

The question about the reality of God, and why He is seemingly so reluctant to give us an absolutely clear sign of His existence (certainly not in some mathematical expression or, as in Sagan's *Contact*, in the digits of pi), brings us full circle back to the discussion in Chap. 1 (Sect. 1.4). There are a number of stories in SF that deal with this issue, spanning the entire spectrum of seriousness. Starting at the low end, with what I think the least serious, is the ham-handed 1951 tale "Second Genesis" by British author Eric Frank Russell (1905–1978). In it a relativistic space traveler who left Earth 3 years ago *in spaceship time* (but 2,030 years ago in Earth time) finally returns home. He finds all those he knew gone (of course), but in fact there is *nobody at all* on Earth because humanity killed itself off with biological experiments gone wrong.

At the moment he realizes that he is irreversibly alone, the last man alive encounters positive proof of God—the space traveler sees His stupendously enormous feet (so, perhaps, I should say this story is 'ham-footed') in the middle of a grassy plain. The last man then falls asleep (sure, isn't that what *you* would do that if you saw two REALLY BIG feet?), and so misses God's sigh of "infinite patience" and His decision that "Nothing for it but to try again." The final words in this story are when Russell really loaded it all on far too much: "He took something from the sleeper's side . . . and breathed into it the breath of life. Leaving the woman to await the man's awakening, the Stranger went away."

Russell did show *some* restraint by naming his traveler Arthur rather than Adam—but at least one author couldn't resist the temptation. In John Brunner's 1956 story "The Windows of Heaven" we read of the first manned landing on the Moon. (God, Himself, doesn't appear in this tale, but His word, as reported in the Bible, does.) Just after touching down on the satellite, there is a huge solar flare, just one step below a minor nova. When the astronaut—who is always referred to by only his last name, Arkwright—returns to Earth, he soon realizes he is the last living creature on the planet. That's because the entire Earth's surface "was changed beyond recognition . . . it was a vision of Inferno." All the oxygen in the atmosphere has been exhausted by the huge fires that roared across the planet, and carbon dioxide is the major atmospheric component.

Arkwright is at first in despair, realizing that all of civilization is gone, and everything is right back to 'the Beginning,' with the air full of CO_2, the oceans near the boiling point, and all the Earth barren of life. After he dies, where will new life come from? The situation is, he thinks, something like it must have been during Noah's flood. But then he discovers a culture slide in his experiment kit (that was to be used to test Moon rocks) has a tiny bacteria colony on it. And so, now smiling at the certainty of what he must do, Arkwright opens the airlock of his ship and, with the culture slide in hand, steps out. Just in case the reader hasn't yet caught the pun of **Ark**wright's name, the tale ends with "And Noah went forth . . . ." (Why the culture slide is necessary in this story is a puzzle to me: after all, Arkwright's own gastrointestinal tract is *full* of bacteria!)

8.2 (Slightly) More Serious Speculations

To *really* be convinced of God's reality, however, what we are *really* talking about is the direct, irrefutable occurrence of a miracle. That is, what we want is a clear violation of one or more of the natural laws of the physical world. We don't have Jesus anymore to rise again from the dead, but there are other possibilities that would work just as well (or so it would seem). For example, what, asked SF editor Lester del Rey of three well-known SF writers, would be the world's reaction to something as dramatic as, say, the planet stopping dead in its orbit for a full day? As he put the challenge to Poul Anderson, Robert Silverberg, and Gordon Dickson (1923–2001), what if "the earth moved not around the Sun, neither did it rotate. And the laws of momentum were confounded." Del Rey asked each to independently write a novella based on that common theme, and then he published the results in the 1972 book *The Day the Sun Stood Still*. All three contributions are quite interesting reading, while following different paths.

In Anderson's "A Chapter of Revelation" there is much discussion of the interwoven nature of politics and religion, and how the principal actors in each field maneuver to gain advantage from the miracle. The story ends with one character wondering why people continually fail to see that the *natural* world, alone, has always been a miracle. (Just think, the only reason for any of us to be alive is because the Earth orbits, at just the right distance, a thermonuclear, gaseous fusion reactor a million miles in diameter, a flaming sphere that converts 4,000,000 t of mass to pure energy *every second*. That *does* sound like something out of a silly SF story, right?) Anderson does make one particularly interesting observation about the appearance of any miracle: "*What* God wrought the miracle, if any did? The Christian God? If so, what version? A medieval one surrounded by saints and angels, a Calvinist one Who plans to cast most of us into perdition, a simpering Positive Thinker, or what?

How do we know it wasn't the Jewish God ... one of the Jewish gods? Or Moslem, or any of a thousand Hindu deities, or what?"[4]

In Silverberg's "Thomas the Proclaimer" we have two characters who represent the two major protagonists in any tale of science and faith: the religious Thomas, and the physicist Gifford ('he wasn't *hostile* to organized religion, he just ignored it'). For Gifford, "the only universal truth there is that Entropy Eventually Wins."[5] This cynicism is matched by, oddly, *fear* on the part of the religious. At one point a woman confronts Thomas about what has happened and, unconvinced of his sincerity at mouthing the usual platitudes of 'Have faith, Pray a lot,' she thinks to herself "He's scared. We're all scared ... and last night the Apocalyptists burned the shopping center."

Others also question if the miracle actually came from Satan, and not God, and numerous, diverse religious cults spring-up everywhere. The Pope says *he* is unsure if God is the author of the miracle, and bored technicians at a satellite relay station transmit a fake radio broadcast from the Almighty. Scientists, led by Gifford, become the new 'authorities' on superstition and pomposity, replacing the Church's priests in that function, but they go too far when they hold a mass burning of holy books, including the Old and New Testaments, the Koran, the Talmud, the Bhagavad-Gita, and more. In response the scientists are attacked by a mob of religious zealots, and Gifford is clubbed to death in a pit full of mud-encrusted Bibles. As the story ends, Thomas, too, dies at the hands of a mob of religious fanatics who think he has failed at his task of Prophet. Nobody would deny that the whole business has been less than satisfactory and, as Silverberg remarks, "Those who begged a Sign from God ... would be content now only with God's renewed and prolonged absence."

In Dickson's "Things Which are Caesar's" the stopping of the Earth is ultimately dismissed as illusionary, as nothing but the result of mass hysteria. One character, who has evidently been on more than one drug-induced 'trip,' claims he has often seen the same purported miracle, and others that are even more impressive. So, big deal. It is, in fact, pretty nearly impossible to think of just *what* God might do that would convince all of His existence. To repeat the words of C. S. Lewis from Chap. 1 (Sect. 1.2): "If the end of the world appeared in all the literal trappings of the Apocalypse; if the modern materialist

---

[4] In late October 2013 an interesting debate on this very issue erupted on the editorial pages of *The Boston Globe*. It was started by an essay whose author claimed it is blasphemy to use the phrase "Oh my God" as an exclamation of wonder, and that teenagers who insert OMG into their social media messages risk eternal damnation. Soon after, the following reply appeared on the Letters-to-the-Editor page: "Why is ... so concerned about the use of 'Oh my God' by others who do not share her faith? When I use the expression, I am not referring to her deity, but to mine: the Flying Spaghetti Monster."

[5] To give this a blunt translation: You are born, you live, you die, you decompose, and that's all there is. If you don't like that, the Universe doesn't care.

[Lewis' word for a skeptic] saw with his own eyes the heavens rolled up and the great white throne appearing, if he had the sensation of being himself hurled into the Lake of Fire, he would continue forever, in the lake itself, to regard his experience as an illusion and to find the explanation of it in psychoanalysis, or cerebral pathology."

I think Lewis has a pretty good argument with that but, still, it would take a *very* odd person to convince themselves that the events in the 2001 "Hell is the Absence of God" are simply illusions. Written by the computer scientist Ted Chiang (born 1967), this incredible tale justifiably won SF's 2001 Hugo award for best short story of the year. It makes no bones about the reality of God; that's a given. When a person dies, his/her soul can be *seen* going in one of two directions: up to Heaven or down to Hell. What more evidence could you ask for? But, if for some reason that isn't enough for you to believe in God's existence, how about this: angels make regular (if unpredictable) 'visitations' in public view, with great drama, appearing in towering pillars of flame. Wow!

In addition, now and then the ground becomes transparent and you can actually *see* Hell beneath your feet, just as if you were looking through a hole in the floor. Double wow! And sometimes the saved souls of deceased relatives appear—occasionally right in the kitchens of the still living!—accompanied by a golden glow of light. They never say anything, but they always have beatific smiles on their faces. Triple wow! So, naturally, nobody has any doubts at all that God *exists*. The great problem is that of *loving* God, which is not a clear-cut proposition. That's because when angels appear—at which times they often bestow miracle gifts on those who happen to be near-by (cancers are cured, the blind have their eyesight restored, cripples can walk again, and so on)—disaster can result.[6]

This is what happens to Neil, the protagonist in this clever tale. Or, rather, it happens to his wife Sarah, who is eating in a café when an angel visitation occurs. The window she is sitting next to shatters and, cut by a multitude of flying glass fragments, she bleeds to death. After she cries in pain and fear while dying, onlookers finally see her soul ascend to Heaven. There is no puzzle about that, as Sarah was always a devout believer in God and she worshipped Him, and was ever grateful to Him for her wonderful life on earth with Neil.

Before Sarah's death, Neil had always assumed that upon his own death he would go to Hell. That's because he doesn't worship God like Sarah did.

---

[6] The entire story is told in a matter-of-fact, deadpan fashion, with Chiang coming at times very close to being flat-out funny. For example, we are told that, when disaster results from an angel visitation, property damage claims are "excluded by private insurance companies due to the cause." No further elaboration is given but, apparently, this is a sly reference to the typical insurance loophole of not covering 'acts of God.'

Going to Hell doesn't really seem so bad either because, when you look down into it everybody there seems to be pretty normal. Certainly there are no devils with pitchforks tormenting damned souls screaming with pain in eternal fire pits of flaming sulphur, images that have given generations of children bedtime nightmares on the Sunday nights after church. Eternity in Heaven would be incomparably superior to spending it in Hell, of course, but all-in-all Hell doesn't seem so bad to Neil. Indeed, some people whose mates had died and gone to Hell had it easy—they just committed suicide (which of course as a sin insured that they went to Hell, too) and so were happily reunited with their mates. Happy in Hell together, forever. How ironic!

With Sarah in Heaven, however, Neil's 'burning' (no pun intended) desire is to join her *there*. But how can Neil hope to go to Heaven when, even though he *believes* in God, he can't bring himself to *love* the God that has violently taken his wife from him? He can't just 'love God' simply as a means to join Sarah. He has to *love God for Himself* because otherwise he wouldn't be demonstrating *true* devotion. That, Neil can't bring himself to do, and to not love God means that Neil will absolutely go to Hell and so never be with Sarah.

Matters look pretty grim for Neil, but then he realizes there *is* one possible avenue of hope, one maybe-perhaps road to Heaven, allowing him to join Sarah there. All angel visitations are accompanied by a brief burst of Heaven's light. That light is of course infinitely beautiful, and is of such compelling majesty that to see it vanquishes all inability to *love God*. To see it can be dangerous, too, as it is known to blind people. But to Neil it is worth the risk, as all who have seen it have *always* been accepted into Heaven, even if they had lived a sinful life. For example, the soul of a notorious serial rapist and murder who had seen Heaven's light while disposing the body of his final victim was, upon his execution, observed to ascend to Heaven. That naturally didn't seem fair to the family members of his victims but, as the old saying goes, 'it is what it is' and so just had to be accepted. God works in mysterious ways, you know.

Well, unfair it might be but it offers a possible solution to Neil's problem. Despite not *loving God* he could nevertheless join Sarah in Heaven *if* he could just arrange to see Heaven's light. So, he becomes one of numerous so-called *light-seekers*, people who frequent particular locations where the statistical odds of an angel visitation are unusually high (these places are called *Holy Sites*). It would be at such a place that Neil would have the best chance of seeing Heaven's light. And so, indeed, his big chance comes—and he almost blows it: chasing after an angel visitation at a Holy Site in a remote desert location while driving "a pickup truck equipped with aggressively knurled-tires and heavy-duty shock absorbers" to navigate the Holy Site's rough terrain, Neil crashes into a boulder, nicks his left femoral artery, and starts to bleed to death. Who could blame Neil for thinking, at that moment, he had failed?

But then, just before actually dying, he is struck full in the face by a blinding flash of Heaven's light! At that instant Neil truly, sincerely, devotedly, *loves God*. Then, he dies. Alas, even though he has seen Heaven's light, Neil's soul still goes to Hell. Remember, God works in mysterious ways.

Neil's existence in Hell hasn't really been terribly awful, except for knowing he will never again be with Sarah. Even with that depressing knowledge, though, he continues—because he has seen Heaven's light—to deeply *love God* despite also knowing that all those in Hell are beyond God's awareness. God simply does not care about Hell or any of its occupants. God's reaction to Hell is simple: to hell with it! No matter: having seen Heaven's light, Neil *loves God*, even though he is not loved by God in return.

That's the hell of Hell.

On one level "Hell is the Absence of God" can be read as simply a cruel joke by a cruel God, as a modern version of the Old Testament story of Job who is a pawn in a game played by a prideful God and an evil Satan. On a deeper level, the story is a tale that delivers the sobering message that to love God with the expectation of that love being returned is a fundamental error. As Chiang writes near the end of his story, "God is not just, God is not kind, God is not merciful," and yet, to love Him anyway is the *only* true measure of *true* devotion.[7] With that in mind, can you think of any other description for this story more apt than it is a *horror story*, one far more frightening than anything Stephen King has ever written?

Chiang's terrifying take on God *would* explain the ancient question of why a benevolent God allows evil and suffering to exist in the world. One answer was given by the novelist Ayn Rand in her 1957 *Atlas Shrugged*, when she has one character say to another (in a non-religious context), "Contradictions do not exist. Whenever you think that you are facing a contradiction, check your premises. You will find that one of them is wrong."[8] There are just two premises at hand: God exists, God is benevolent. Take your pick on which to deny.

---

[7] When I read these words by Chiang, I was reminded of how SF writer Philip K. Dick described God after recovering (sort of) from his second LSD 'trip': "I perceived Him as a pulsing, furious, throbbing mass of vengeance-seeking authority, demanding an audit like a sort of metaphysical IRS agent." From the Chronology of Dick's life, in *VALIS and Later Novels*, The American Library 2009, p. 830.

[8] In Chap. 7 ("The Exploiters and the Exploited") of Part 1.

8.3 The End of It All

To end this chapter—and the book, too—I'll finish with a discussion of Asimov's 1979 story "The Last Answer," as I promised back in Chap. 5 (Sect. 5.3). It is, I think, one of the very best stories he ever wrote, and I suspect it illustrates his personal beliefs (and even hopes) concerning God and the hereafter. Asimov was never shy in expressing his belief that nothing lies beyond the grave and yet, if by some chance he should be wrong, then his story shows what *he* would hope to be doing in his new existence. (As you'll see, that didn't include lounging in eternal bliss on a cloud, listening to harp music!)

As we begin reading we watch an atheistic physicist drop dead in his laboratory from a heart attack. Much to his surprise, his consciousness continues to think, and then to begin interacting with an entity he calls The Voice. When he asks (via some sort of direct transfer mechanism of thoughts) the obvious question—"Are you God?"—the reply is enigmatic at best: "There is no answer I can give that you would comprehend. I *am*[9] . . . which is all that I can say . . ."

When the physicist tries to pursue the religious theme, by asking "And what am I? A soul?" he gets an answer that every physicist can appreciate (although theologians might be a bit less so inclined): "You may call yourself a soul if that pleases you, but what you are is a nexus of electromagnetic forces, so arranged that all the interconnections and interrelationships are exactly imitative of those in your brain in your [prior] existence—down to the smallest detail . . . It still seems to you that you are you."

After this 'getting to know you' prelude, there then follows the steady extraction by the physicist of just what The Voice intends. The physicist learns that his continued awareness, even after death, is not unique but still is quite rare. He has been selected, along with others from all the intelligent species in the entire universe, not for any spiritual reasons but strictly because of his ability for high-level thinking. The Voice is as old as the universe which it made ("it is my invention, my construction"), and so The Voice may even be eternal, and its knowledge infinite. But that doesn't mean it knows *everything*.

---

[9] See note 8 in Chap. 5 again.

And matters are even more complicated than that. As The Voice explains, "Even if I knew everything, I could not know that I know everything . . . I have existed eternally,[10] but what does that mean? It means I cannot remember having come into existence. If I could, I would not have existed eternally. If I cannot remember having come into existence, then there is at least one thing—the nature of my coming into existence—that I do not know." The Voice gives the physicist a mathematical illustration of what it means to both have infinite knowledge and yet to still have gaps in that knowledge. The Voice could know, for example, the infinity of the *even* integers, while remaining ignorant of the infinity of the *odd* integers.

When the physicist observes that knowledge of the even integers implies knowledge of the odd ones, too ("if you divide every even integer . . . by two, you will get another infinite series which [contains] within it the infinite series of the odd integers"), he is told that The Voice is greatly pleased. That's because the physicist's observation has demonstrated his ability to successfully carryout The Voice's plan, which it at last reveals: "I constructed the Universe in order to have more facts to deal with. I inserted the uncertainty principle, entropy, and other randomization factors to make the whole not instantly obvious. It has worked well for it has amused me throughout its entire existence. [However] I found I could not predict the next interesting piece of knowledge gained, where it would come from, by what means derived."

---

[10] The eternal nature of God is the theme of William Blake's 1794 watercolor "Ancient of Days," the name for God in Chap. 7 of the Bible's Book of Daniel. (A similar usage occurs in holy works of Judaism, Hinduism, and Buddhism.) In Blake's painting God's hand is shown holding a compass to 'measure the universe,' and this coupling of a mathematical instrument with faith mirrors the appearance of religion in SF.

Blake's "Ancient of Days" (British Museum)

In other words, The Voice—clearly 'God'—made the universe for the entertainment(!) of continually learning ever more. God, the perpetual student. But in that noble enterprise even the boundless powers of The Voice are not enough, and it needs help! (As the story develops, however, we learn that The Voice is not being completely transparent with the physicist, and it is not just more knowledge it is after, but one *particular* bit of knowledge. More on that, soon.)

When the physicist expresses skepticism that *he*, until recently a mere mortal, can really find new knowledge that has escaped The Voice, he is assured that his "success is certain, since you will be engaged eternally." To that, the physicist objects. He tells The Voice that, for him, the joy in pursuing knowledge comes from discovery *in a finite lifetime* that only he could do. To do what The Voice asks would give no joy as, after all, The Voice could itself do whatever the physicist might accomplish. Further, when alive on Earth the physicist did his work with the potential praise of his fellow physicists as a reward. By contrast, all that The Voice offers is its 'amusement.' As he tells

The Voice, "there is no credit or satisfaction in accomplishment when I have all eternity to do it in."

And so the physicist says he *won't* do it.

The Voice admits it doesn't want to force the physicist to think, and in fact doesn't have to because, really, what else is the physicist going to do in his new existence *but* think? To that the physicist agrees, but replies that the only thing he will think about is how to "disrupt the nexus of me. *That* would amuse me." In reply, The Voice is surprisingly agreeable, saying in effect 'feel free to do so.' That's because, as it explains, "if you succeed in this suicide attempt you will have accomplished nothing, for I would instantly reconstruct you and in such a way as to make your method of suicide impossible. . . . It could be an interesting game, but you will nevertheless exist eternally. It is my will."

This might appear to be the end of any possible resistance from the physicist, but he has one last card to play. Rather than trying to destroy himself, he tells The Voice "I will set as my goal the humiliation of you. I will think of something you have not only never thought of but never could think of. I will think of the last answer, beyond which there is no knowledge further." To that The Voice disagrees, pointing out that "There may be things I have not yet troubled to know [but] there cannot be anything I cannot know."

That seems reasonable, but the physicist counters with this: "You cannot know your beginning. You have said so. Therefore you cannot know your end. Very well, then. That will be my purpose and that will be the last answer. I will not destroy myself. I will destroy *you*." You might think The Voice would be outraged at such a threat, but not so. Indeed, The Voice is pleased! It is then that the physicist finally understands The Voice's real goal: "For what could any Entity, conscious of eternal existence, want—but an end?" (Read, once more, Swinburne's poem "The Garden of Proserpine" in Chap. 6.)

In Asimov's tale 'God' is bored, and the only escape is 'death.' I'm not convinced, however, that Asimov has quite made his case, logically. Why does The Voice need the physicist, since anything the once 'mere mortal' could do The Voice could do for itself just as well? If The Voice thinks long enough (and it has eternity available) then it should be able to figure-out how to end its own existence, all by itself. Or have I missed something?

If you have an answer, write and tell me. But, until then, let this be the end (but *not* of *everything*—at least not yet).

Appendix 1: Matching Wits with 'God'[1]

If you think a problem that asks you to accept the possibility of the situation suggested by the above title is simply silly, that it poses a situation nobody with any serious intent would suggest (outside of theology, of course), you are wrong. In his *God and Golem, Inc.*, cybernetics guru Norbert Wiener (see the opening paragraph of Chap. 4) writes that "to play a game with an omnipotent, omniscient God is the act of a fool." What we'll do here, however, is not quite what Wiener had in mind.

In Chap. 1 I mentioned a book by the political scientist Steven Brams, in which he used two-person game theory to study the outcomes of an ordinary person interacting with an 'opponent' that possess the attributes of omniscience, omnipotence, immortality, and incomprehensibility. That is, he studied the interactions of a human 'playing against' what he called a 'superior being'—or, if you wish, against God. In 1950 the mathematicians Merill M. Flood (1908–1991) and Melvin Dresher (1911–1992), while working at The RAND Corporation in Santa Monica, California (an Air Force think tank), jointly created a game theory puzzle question that makes this very suggestion, and it has bedeviled analysts ever since. I'll first describe it in its best-known, non-probabilistic form, and then again in the form that gives this appendix its title (and in which a small amount of *very* elementary probability makes an appearance). There is still no known analysis that satisfies everybody. Indeed, I suspect God, Himself, may be scratching His head!

The best-known version of the original Flood/Dresher puzzle is called the "Prisoner's Dilemma," a name given to it by Albert W. Tucker (1905–1995), a

---

[1] This appendix, in very slightly different form, originally appeared in my book *Will You Be Alive Ten Years From Now? and numerous other curious questions in probability theory*, Princeton University Press 2014. I thank PUP for permission to reprint.

P.J. Nahin, *Holy Sci-Fi!*, Science and Fiction,
DOI 10.1007/978-1-4939-0618-5, © Springer Science+Business Media New York 2014

Princeton University mathematician. Imagine that you and another person have been arrested and each of you have been charged with two crimes, one serious and the other not so serious. You've both vigorously claimed innocence, but are now being held in separate cells awaiting trial. There is no communication possible between the two of you. Then, just before the trial is to start, the prosecuting attorney from the DA's Office shows-up in your cell with the following offer.

There is sufficient circumstantial evidence to convict both of you of the not so serious charge, enough to get each of you a year in prison even if neither of you confess. But, if *you* will confess then the other person will be convicted of the more serious charge and get 10 years in prison *and you will be set free.* When you ask if the other person is getting the same offer, the answer is 'yes' and, further, when you ask what happens if both of you confess the reply is that then both of you will get 5 years in prison. The puzzle question is now obvious: what should your decision be, to confess or not?

To help keep all the conditions clear in your mind, the following table of your various potential fates should help:

Actions	Other person confesses	Other person doesn't confess
You confess	you get 5 years in prison	you go free
You don't confess	you get 10 years in prison	you get 1 year in prison

To make your decision, you might use the following standard game theory reasoning. The other person is either going to confess or not. It's going to be one or the other, and which it is has nothing to do with anything *you* can control. So, suppose he/she does confess. If you confess you get 5 years, and if you don't confess you get 10 years. Clearly, you should confess *if he/she confesses.* But suppose he/she doesn't confess. If you confess you go free, and if you don't confess you get 1 year. Clearly, you should confess *if he/she doesn't confess.* That is, you should confess *no matter what* the other person decides to do. For you to confess is said to be (in game theory lingo) the *dominant decision strategy.*

But here's the rub. The other person can obviously go through exactly the same reasoning process as you've just done, to conclude that his/her choice is also dictated by the dominant decision strategy of confessing. The end result is that you *both* confess and so you *both* get 5 years in prison! The 'paradox' is that perfectly rational reasoning, by each of you, has resulted in a non-optimal solution because if you both had simply kept quiet and said nothing, then you both would have gotten the much less severe sentence of 1 year in prison. Philosophers have argued (for *decades*) over whether this is really a paradox or merely 'surprising,' and the literature on the problem had, even years ago, grown to a point where nobody could possibly read it all in less than that 10 year prison sentence. And it continues to grow ever more voluminous even as I write.

It was while thinking about the "Prisoner's Dilemma" in 1960 that William Newcomb (1927–1999), a theoretical physicist at the Lawrence Radiation Laboratory—now the Lawrence Livermore National Laboratory (LLNL)—in California, created an even more perplexing puzzle. Newcomb's problem (now called *Newcomb's Paradox*) was formulated to help him explore the "Prisoner's Dilemma," and it is now generally believed that Newcomb's Paradox is a generalization containing the Prisoner's Dilemma as a special case.

Curiously, Newcomb himself never published anything about his puzzle, but instead it first appeared in print in a 1969 paper by the Harvard philosopher Robert Nozick (1938–2002). The puzzle had been circulating via word-of-mouth in the academic community, but Nozick decided it needed a much wider audience. But what really brought Newcomb's puzzle world-wide fame was when it appeared in the July 1973 "Mathematical Games" column of *Scientific American* (with a follow-up column in the March 1974 issue), written by the well-known popular math essayist Martin Gardner (1910–2010). So, here's Newcomb's Paradox.

Imagine that you are approached by an intelligent entity that has a finite but lengthy history of predicting human behavior with unfailing (so far) accuracy. It has, to date, never been wrong. You may think of this entity as (using Gardner's examples) a "superior intelligence from another planet, or a super-computer capable of probing your brain and making highly accurate predictions about your decisions." Or, if you like, you can think of the entity as being God. This entity makes the following presentation to you.

A week ago, the entity tells you, it predicted what you would do in the next few moments about the contents of those two mysterious boxes you've been wondering about that are sitting on a table in front of you. The boxes are labeled B1 and B2, and you can either take the contents of both boxes, or the contents of box B2 only. The choice is entirely yours. B1 has a glass top, and you can *see* that the entity put $1,000 in that box. B2 has an opaque top, and you can't see what, if anything, is in it. The entity, however, tells you that it put nothing in B2 *if* last week it predicted you would take the contents of both boxes, or it put $1,000,000 in B2 *if* last week it predicted you would take the contents of only B2.

So, what's your decision? Take both boxes, or just box B2 alone? The reason why this situation is called a *paradox* is because there are seemingly two quite different (but each clearly rational) ways to argue about what you should do—the two ways, however, lead to *opposite* conclusions! The first line of reasoning is similar to the one we used in the Prisoner's Dilemma, in that it is a dominance argument. As we did there, let's make a table of the various potential outcomes as a function of what you decide and what the entity predicted:

Actions	Entity predicted you'll take both boxes	Entity predicted you'll take only box B2
you take both boxes	you get $1,000	you get $1,001,000
you take only box B2	you get nothing	you get $1,000,000

Now, the entity (you reason) made its prediction *a week ago* and, based on that decision *then*, either did or didn't put $1,000,000 in B2. Whatever it did is a done deal and can't be changed by what you decide *now*. So, looking at the above table, it's clear that you have the dominant strategy of taking the contents of both boxes, as $1,000 is greater than nothing (the entity predicted you'd take both boxes), and $1,001,000 is greater than $1,000,000 (the entity predicted you'd take only B2).

Okay, that all makes sense to a lot of people, maybe you, too. But—there is another, probabilistic argument that leads to the opposite conclusion. It goes like this. We don't *know* that the entity is absolutely infallible. Yes, it's true that it hasn't been wrong yet, but its track record *is* finite. So, let's say it has probability p of being correct and, since it has always been right up to now, it is almost certain that p is pretty close to 1 (but we don't know that it *is* 1). So, for now it's p. Now, in decision theory there is, besides the dominant strategy principle, another equally respected principle called the *expected-utility* strategy, in which you decide what to do by maximizing the expected utility that results from your choice. The utility of an outcome is simply the product of the probability of the outcome by the value of the outcome, and the expected utility is the sum of all the individual utilities.

So, suppose you decide to take both boxes. The entity would have predicted (correctly) that you would do that with probability p, and with probability 1—p it would have predicted (incorrectly) that you'd take only B2. So, the expected utility resulting from the choice of taking both boxes is

$$U_{both} = 1,000p + 1,001,000(1 - p) = 1,001,000 - 1,000,000p.$$

Next, suppose you decide to take only B2. The entity would have predicted (correctly) that you would do that with probability p, and with probability 1—p it would have predicted (incorrectly) that you'd take both boxes. So, the expected utility resulting from the choice of taking only B2 is

$$U_{B2} = 1,000,000p + 0(1 - p) = 1,000,000p.$$

Notice that as $p \to 1$ we have $U_{both} \to 1,000$ while $U_{B2} \to 1,000,000$.

So, the expected utility principle says you should decide to take only B2 *if* the entity is almost always correct. In fact, we can *very* loosely interpret what 'almost' means since as long as $p > 0.5005$ (the entity simply flips an almost fair coin to make its prediction!) we have $U_{B2} > U_{both}$ and the expected utility principle says you should take only B2.

I think you can now clearly see the 'paradox' in Newcomb's Paradox. Two valid arguments, each eminent examples of rational reasoning, have led to exactly opposite conclusions. As Professor Nozick wrote in his 1969 paper,

> "I have put this problem to a large number of people, both friends and students in class. To almost everyone it is perfectly clear and obvious what should be done. The difficulty is that these people seem to divide almost evenly on the problem with large numbers thinking that the opposite half is just being silly. Given two such compelling, opposing arguments, it will not do to rest content with one's belief that one knows what to do. Nor will it do to just repeat one of the arguments loudly and slowly. One must also disarm the opposing argument; explain away its force while showing it due respect."

Well, logicians, philosophers, mathematicians, physicists, and just plain folks have been trying to do that over the more than 40 years since Nozick wrote, and the noise and confusion continues to this day. What, you might wonder, did the *creator* of this puzzle think should be the choice? In a recent contribution,[2] the physicist and SF writer Gregory Benford (who once shared an office with Newcomb at LLNL and often discussed the problem with him, long before it became famous) revealed that when he asked Newcomb that very question the reply was a resigned 'I would just take B2; why fight a God-like being?' I read that as meaning Newcomb, too, was as stumped by his own puzzle as has been everyone else![3]

This intellectual conundrum reminded Martin Gardner of one of the amusing little poetic jottings of the Danish scientist Piet Hein (1905–1996):

> "A bit beyond perception's reach
> I sometimes believe I see
> That life is two locked boxes, each
> Containing the other's key."

That might well be a good summary of most of the issues discussed in this book!

---

[2] David H. Wolpert and Gregory Benford, "The Lesson of Newcomb's Paradox," *Synthese* (Online First), March 16, 2011. There are a lot of references in this paper to the vast literature on the problem.
[3] One writer who directly associated Newcomb's 'God-like being' *with* God was Dennis M. Ahern: see his paper "Foreknowledge: Nelson Pike and Newcomb's Problem," *Religious Studies*, December 1979, pp. 475-490. Ahern was at the time a philosophy professor at the University of Maryland, and Nelson Pike (1930-2010) was a philosophy professor at the University of California at Irvine. Ahern was responding in part to Pike's paper "Divine Omniscience and Voluntary Action," *Philosophical Review*, January 1965, pp. 27-46.

Appendix 2: (fantasy)

"Some Things Just *Have* To Be Done By Hand!"

I wrote this short-short in 1978, and submitted it to *Analog Science Fiction* as a possibility for the magazine's 'way-out' page. Called "Probability Zero," the name of that page is intended to let readers know that both the author and the editor (at that time, Ben Bova, *Analog*'s second editor after John Campbell's death in 1971) know it isn't science fiction but rather is *fantasy*. Bova bought the piece but, before he could print it, he left *Analog* to become the first fiction editor at a new glossy science fact and fiction magazine called *Omni* (started in 1978 by *Penthouse* founder, the late Bob Guccione). The new editor at *Analog*, Stanley Schmidt (who only recently retired in 2012) decided to use it in the 1981 *Analog Yearbook 2*, and that's where and when it finally appeared. So, here it is, in very slightly altered form, with God, Himself, as the central character.

****

The Most Important Entity rubbed His temples in fatigue. There was just so damned much crap to put up with nowadays. The personnel paperwork was nearly overwhelming, even for a being with omnipotent powers. And a work force faced with zero turnover had a first-class morale problem. The younger ones knew there was no hope for advancement by the once-usual routes of death, retirement, or resignation. None of those events ever happened—here.

The telephone rang, and He answered in weary relief at the distraction. "Yes?"

"Sorry to bother you, Sir, but the main computers have a backlog in the RANDOM QUEUE for ten to the 183rd power decisions. Can you please service those requests right now?"

P.J. Nahin, *Holy Sci-Fi!*, Science and Fiction,
DOI 10.1007/978-1-4939-0618-5, © Springer Science+Business Media New York 2014

"Damn, are those bloody scientists on Earth doing their quantum experiments again!? You'd think they'd understand the Uncertainty Principle after all these years. Well, what is it now, an electron beam through a diffraction grating, or is somebody trying to locate an atom with zero error?"

"Both, and more, Sir. Those guys are really getting busy down there. Why, just as we've been talking here, the RQ has picked up ten to the 179th power more requests!"

The main computers couldn't be allowed to overflow. Once, two or three thousand years ago (in Earth time), they had been unattended for several days (in His time), and the RQ had clogged up tight with ignored decision requests for determining the outcomes of random events. The resulting massive computer system crash had caused entire centuries (in Earth time) of strange, abnormal violations in His Laws of Natural Phenomena. It had been the time of magic on Earth, and the new wizards, sorcerers, and magicians had used it to their advantage in proclaiming themselves to be all-powerful. It couldn't be allowed to happen again!

"All right, all right, hold your feathers smooth. Hang on for a moment." He put His caller on hold, and pulled open the desk drawer next to His perfect left foot. Inside was a pure diamond crystal box, containing two ruby cubes of ultimate clarity. The dots on the cube faces were precise circles of gold. The cubes were perfectly balanced, of course, as it was impossible for anything unfair to exist—here. Taking the cubes in His mighty hand, He established a mind-link with the input-output data lines to the main computers. Faster than imaginable (or even possible by ordinary laws, but for Him very little was impossible) the cubes tumbled in His quivering hand. The whole thing was over in just a few wing beats. He dropped the cubes, now so hot they glowed in the gamma-ray region of the spectrum, back into their crystal box, and shoved the drawer shut with a kick from His perfect left foot.

"Okay, the main computers cleaned up?"

"Yes Sir, the RANDOM QUEUE is empty!"

"Excellent—now please don't call again for at least another day. Meanwhile, you and your colleagues might busy yourselves with finding a way to speed up the automatic software random number generator. I find this business of hand-generation to be increasingly inconvenient. Good-bye."

As He hung-up, He thought of what Albert Einstein, one of the better Earth scientists, had once said: "God doesn't play dice with the Cosmos."

"Hummph," He grunted in disgust to Himself, "just what the Hell did *he* know about it?"

****

"The Next Time Around"

This second fantasy short-short, dealing (I suspect most people would say) in a pretty irreverent way with reincarnation, was written in a single sitting in 1979, as a break from a late-night session of exam grading. I was already sort of mentally unbalanced when I started it, from too much bleary-eyed reading of equations scrawled in dull pencil, and perhaps the story reflects that. Nevertheless, I had a lot of fun writing it and so I sent it to some of the big name SF magazines: *Isaac Asimov's*, *Omni*, *Analog*, even *Playboy*. No takers, but Stan Schmidt at *Analog* wrote back to say that while he liked it, it was just a bit too much "Twilight Zone" for *Analog's* readers. Well, now, *that's* a thought, I remember thinking, and so off it went in the next mail to T. E. D. Klein, the editor (from 1981 to 1985) of *Rod Serling's The Twilight Zone Magazine*. Much to my pleasure (and relief), he bought it. In a funny little note that he wrote in his offer (I still have it), he told me he was in the same sort of situation I had been when I wrote: "It's past 4 a.m. now, I'm still in the office and—feeling unusually efficient—I'm taking the risk of offering you, right now, the enclosed contract . . . ." The story appeared in the August 1981 issue of the magazine, with Klein's editorial lead-in reading "When you're speeding down the highway at 70 m.p.h., what better time to think about life . . . and death?"

****

Rollo Adams pulled out of the motel parking lot just before dawn. It was best to hit the smooth, hard pavement of the superhighway while the air was still cold and the concrete slightly wet with dew. The souped-up convertible accelerated quickly and smoothly to 70, and Adams settled in for the last long day of cross-country travel.

As he watched the flat emptiness of the Arizona desert flash by, he felt the wind blow over his tiny bald spot. The old carcass sure did ache! His left side still hurt where the vandal had kicked him yesterday. If somebody hadn't come along just then, the bastard probably would have stuck him with a knife. Damn him! He tried to forget the discomfort by thinking of his destination—the romantic waterfront of San Diego Bay.

The car backfired once, and he wished he had a tachometer so he'd know how many RPMs he was pulling.

The miles pounded by, and Adams—who, as usual in the morning, had felt worn-out, down, deflated—began to feel a good deal better. The heat of the drive smoothed his stiffness away, and his grip on things was becoming firm again. Thank God the episodes of nausea and dizziness were getting less frequent and severe! For a while he'd been wandering, slipping a bit, but he was gradually gaining experience, and he could handle the pressures of his new life better now.

And what pressures! At first he'd been almost paranoiac about it, always on the watch for danger, never knowing when he might get drilled full of holes! He'd watched all the usual cops and robbers shows on television, sure, but that was just recycled Hollywood fantasy. Now, he *really* knew what it meant to be on the run for the rest of his life.

But the thick steel belts he wore around his vital areas reassured him. With body armor like that, it'd take a mighty big slug to rip *him* open! He had a lot of miles of experience on him now. He'd survived some pretty rough banging around these last few weeks, and had learned how well he could bounce back. He'd been pleasantly surprised. He knew he wouldn't blow and lose control. He was tougher, more resilient, than he'd thought.

The road ahead was empty, a ribbon running long and straight to the horizon, and so he let his thoughts drift back to last month, when he'd experienced the most traumatic event in his life. Man, the only creature on earth to be aware of the inevitability of his own death, still learns to cope with it. But it's one thing to read of the passing of a stranger, or even of one casually known; it's another when death strikes closer to home and snatches away your wife.

A giant wave of loss swept over Adams as he thought of Sally. God, how he missed her! He sighed quietly to himself as he recalled how the two of them had often joked about what came after death. Crazy things, like coming back as some other person. Sally had always said she wanted to return as a lizard and bake all day on a rock in the sun; she had never liked the cold winters in New England. Adams had chuckled at the thought of his elegant wife sitting on a rock eating flies. When he'd mentioned this to her, she'd frowned momentarily and then declared that it didn't matter, because once she was a lizard she'd *like* flies!

Half hypnotized by the rhythmic undulation of the road surface, and with nothing else to distract him, Adams continued to remember. He recalled how Sally had laughed at him when he'd suggested that she might come back all right, but not as a lizard. Maybe she'd come back as the rock! That would be okay, too, she'd replied, tears of silliness running down her face, just as long as she could roast in the sun!

Rollo Adams prayed with all his will that his wife had been granted her request. She had been so young and beautiful to die in that plane crash; he hoped that her wish had come true. And maybe it had! After all, weren't there religions that said you came back as a higher form if you'd led a virtuous life, and a lower one if you hadn't been so good? Who was to say if a lizard, or even a rock, was higher or lower than man?

Perhaps things had worked out for her; perhaps not. They certainly hadn't for him. Maybe those three or four one-night flings a few years ago were the cause of *his* fate. He'd worshipped Sally, and those few moments of weakness still shamed him. He felt the urge to weep, and almost came undone right there. But then his new strength saved him. He had learned, over the past few weeks, to hold everything inside. To let it all out now would be disastrous.

Rollo Adams, dead in the same crash as Sally, roared down the highway. Instantly responsive to the rear-axle, high-torque differential shaft that spun him, he gripped his sporty rear magnesium rim, dug his zigzag slip-proof treads into the road, and felt the pavement rush past beneath him. The road stretched ahead in the hot sun, and San Diego beckoned.

****

I wrote "The Next Time Around" as simply a light-hearted take on resurrection. At the time I had not yet read "Riverworld," a story published 12 years earlier (January 1966) in *Galaxy Science Fiction* magazine by Philip José Farmer (1918–2009). "Riverworld" is a much more serious treatment of resurrection (it's difficult to think of one *less* serious!), opening with Tom Mix (a real-life cowboy movie actor who was killed in a 1940 auto crash in Arizona) in a boat, fighting for his life. Mix, along with 25 billion other humans from all across the ages, was resurrected 5 years earlier on "All Souls' Day" by some mysterious process. All were scattered along the shores of a river ten million miles along, and so we are clearly not on Earth. The only thing that seems evident is that they are all in a place "built by sentient beings," but whether that includes God or not is left unanswered.[4]

One of Mix's companions is slowly revealed to be Jesus who, understandably shocked by how different his resurrected life is from what he expected it to be when he hung on the Cross, has renounced his religion. When Mix, near the end of the story, is about to be burned at the stake by a resurrected but unrepentant fifteenth century Inquisitor, Jesus tells Mix "There was a time when I might have rid you of your pain ... But no more. You have to have faith—and now I do not have it." This is far grimmer stuff than is "The Next Time Around"!

---

[4] This question was eventually answered in Farmer's 1971 novel *To Your Scattered Bodies Go*, which is a vast expansion of the 1966 short story. It wasn't God who created this astonishing world, but rather a race of superior beings. Farmer's fictional creation might seem to be pretty amazing, but at least one reviewer was not impressed: see Franz Rottensteiner, "Playing Around with Creation: Philip José Farmer," *Science-Fiction Studies*, Autumn 1973, pp. 94-98.

Appendix 3: "A Father's Gift"

I've included this third story of mine for two reasons. First, of course, as an example of 'religious' *science fiction* as opposed to *fantasy* (see the first two appendices), but also because it illustrates how even a 'non-believer' (see my opening and closing comments in Chap. 1) should not be constrained by personal bias when writing SF. I write this because the story imagines the scientific conversion of a skeptic (that's me, I admit it!) concerning the divine origin of Jesus.

The story has the following history. It stars an archaeology professor who succeeds beyond his wildest dreams while on a quest for a religious artifact. Sound like an obvious rip-off of Indiana Jones? I hasten to remind you that the first Indy film was released in 1981, while the story was written in early 1979 and appeared in print in the August 1980 issue of *Omni*, a year *before* the film. Years after, in October 1993 and both a movie sequel and a prequel later, I learned that Steven Spielberg and Harrison Ford had agreed to do a fourth installment of the Indy saga as soon as a suitable script could be developed. So I wrote to my then agent in Los Angeles to suggest that he approach Spielberg with "A Father's Gift" as a starting point. After all, the earlier films had told of Indy seeking the Ark of the Covenant and the Holy Grail—the tomb of Christ seemed a logical 'sequel quest' for a religious artifact.

Alas, my agent soon wrote back to say I was too late, that Spielberg had already settled on a concept (which, however, didn't appear in theaters until fifteen—*fifteen!*—years later; who says Hollywood could outrun a snail?). I did later get a couple of movie-deal nibbles, but about 99 % (or more) of Hollywood film nibbles go nowhere, and mine were no exception. But I still think the story could be expanded to make a great adventure film. (*Of course* I'd think that, what author wouldn't?)

As the *Omni* fiction editor (Ben Bova, recently arrived from the editorship of *Analog*) headlined the story in the magazine: *"It was the greatest discovery of*

P.J. Nahin, *Holy Sci-Fi!*, Science and Fiction,
DOI 10.1007/978-1-4939-0618-5, © Springer Science+Business Media New York 2014

the ages: all he had to do was open the coffin." And then, when just a couple of years after the story first appeared I read of how Pope Paul VI had, in 1968, officially endorsed the 'discovery' of the remains of St. Peter (not Jesus, of course, but pretty close), well, in the immortal words of baseball legend Yogi Berra, it was *déjà vu* all over again![5] Since the relationship between the Church and the issue of religious relics—the 'discovery' of which became a big business in the Middle Ages—has always been one bordering on embarrassment (consider, for example, the well-known case of the Shroud of Turin[6]), this was no small announcement. In any case, what follows in "A Father's Gift" is a tale of the greatest possible—on Earth, at least—religious find of all.

****

I have found Christ. No, no, don't say, "Oh, one of *those* people!" Please, hear me out. I'm no zealous religious convert, no fanatic, not even a fallen politician seeking public absolution for misdeeds in office. I'm a hard-science computer archaeologist on the staff of a well-known American university, specializing in the analysis of X-ray axial tomography of the mummies of Egyptian pharaohs. So when I say I've found Christ, I mean I've *found* him.

And something else.

I've never been a devout man. That's led to some interesting discussions over the years with my brother, Jack, who's an associate professor of ancient Middle Eastern languages at Georgetown University. A scholarly Jesuit, Jack had long ago kindled in me a fascination for Jesus Christ the teacher. Can anyone doubt what a truly remarkable man He must have been? But I've never been able to accept the Church's dogma that He was the Son of God, the Savior here on Earth as the result of the Virgin Birth. And who, through the Crucifixion, suffered for the sins of all men. Up to now, I haven't been able to believe, that is. Now—well, let's just say I'm not so sure anymore.

We actually know so little about the life of Jesus, with what we do have coming only from the somewhat confusing, contradictory four Gospels. We do, however, have a fairly good idea of the political times. It was the reign of the murderous King Herod (of whom Caesar Augustus once said he'd rather be a pig than a child in the House of Herod!), the Jews were oppressed by Rome, and the Children of God were eager for the coming of the Messiah long predicted in the Old Testament. The times were ready for a Savior, and Jesus Christ was the right man in the right place at the right time.

---

[5] John Evangelist Walsh, *The Bones of St. Peter*, Doubleday 1982.
[6] Ian Wilson, *The Mysterious Shroud*, Doubleday 1986.

There is little dispute by scholars that Jesus was absolutely certain of His role. His life was no fraud, no shameful act of a charlatan. No actor could have suffered as He did. Some say He left us with His image on the Shroud of Turin and nothing else. They say that He died and disappeared forever.

Or did He?

A number of respected Bible scholars (a minority, yes, but still a significant number) have questioned the traditional description of the death of Jesus, primarily because of their skepticism about the Resurrection. I must admit, that has always been the stumbling block for my willingness to believe, too. Since I am an Egyptologist interested in funerary procedures, the death of Christ has fascinated me for years. It had always seemed to me that there just *had* to be an alternative explanation for what *really* happened. And I was right.

It all started some years ago. While in Cairo at the Egyptian Museum, I was studying dental X-rays of their collection of royal mummies as a sabbatical research project. It was there that I met a brilliant, intense man named Gamal el-Zam, now deceased. He was a professor of philosophy at the University of Alexandria and was also on sabbatical leave at the museum's antiquities department. I became friendly with Professor el-Zam, and soon we were discussing our various research activities. Somehow the discussion got around to my interest in Christ and my conviction that His death was still a mystery, no matter what the Bible may actually say on the matter. I recall he stared quizzically at me for a few moments, and I could see that he was debating in his mind whether or not to pursue it. He must have sensed the depth of my interest, because he plunged on.

"So, my friend, you are a doubter, are you? Good! Possibly, then, you will find some papers I have curious reading. There is great uncertainty about their veracity, as I believe they are actually a transcript of a lost part of the Apocrypha in St. Jerome's Vulgate. The Catholic Church rejects them. But who is to say—if you are as interested in pursuing the details of Christ's death as you seem, then maybe they will be of help. But I warn you, you may be getting into more than you bargain for."

What he actually gave me weren't the ancient manuscripts themselves. The original documents have long been lost, and it was photocopies of these rediscovered manuscripts that Professor el-Zam had, including the papers he suspected would interest me.

After Christ was taken down from the Cross on the hill of Golgotha outside Jerusalem, His body was, according to the evangelists, taken by the wealthy Joseph of Arimathea to a nearby tomb cut in rock. After that the body disappears, and Gospel records become what unbelievers call myth, with the story of the Resurrection 3 days later, and the ascent into Heaven after 40 additional days. Of course, the Gospels are shaky on this point, too, since Luke also says Christ went up to Heaven on Easter Day, well before the 40 days were ended.

That there are 40 days between the Resurrection and the Ascension has always fascinated me, because Genesis itself mentions this as the usual time required for embalming. Could it be that the followers of Jesus spirited His body away from the rock tomb to prevent its defilement by the Romans, who might have buried it in a common criminal's grave? Could it be that a select adherent embalmed the body of the Messiah and then secretly buried it?

The long-lost records given me by Professor el-Zam gave me the answers. Among them was a letter from Joseph of Arimathea to a man named Tertullian, apparently a close friend. First swearing him to secrecy, Joseph then describes the real fate of Jesus. The letter was in the ancient dialect of the common masses, Galilean Aramaic, which I could read only with great difficulty. Making an exact hand copy of the letter, but carefully deleting all references to Jesus by name, I sent it to my brother, Jack, over in America. My wait for his reply was agonizing. It came 3 weeks later:

Greetings to my beloved, but unrepentant brother! The strange text (*where did you find it?*) you recently sent was most challenging. I enjoyed the mental exercise, but it has left me somewhat perplexed. When you return to the States, I want to have a long talk with you about it. But, in answer to your request, here is my version of the original:

> "And we bound his body in fresh linen and sealed his wounds. To secret its final fate, it was taken by night to a faithful follower, also a practitioner of the ancient art of preservation of the Egyptians. There it was purified, covered in soft lead sheets, wrapped in bandages, and sealed into a box of the Pharaoh. Transported overland to the Nile River, it was then sent by boat to the south, to the Temple of the Four Kings. There it was buried, safe at last from the Romans."

As I read these words—words I had translated crudely myself but now was sure were right—I could barely contain my excitement. Jesus had been embalmed, and His body was mummified and then secretly shipped to Egypt and buried. But what was even more incredible was that I also sure the coffin was no longer at its original site, I was also sure I knew where it was. Right where I sat, in the Egyptian Museum itself!

But I wasn't the only one to see these new documents. Had anyone else reached the same conclusion as I? A conversation I had with Professor el-Zam, the day after getting Jack's translation, put my concern to rest. He was then about to return to Alexandria, and he inquired about my reaction to the photocopies he'd given me.

"So, my friend, what do you think about those ancient documents, now that you've had the opportunity to examine them?"

I answered carefully. Just how much did the professor know? "Gamal, they're fascinating. But one letter among them is *most* fascinating; the letter from Joseph of Arimathea to Tertullian, concerning the events after the Crucifixion. Have you read that one, Gamal?"

"Oh, yes. Interesting to *you*, no doubt, because of your curious fixation on the death of Christ. But surely it cannot be authentic. After all, the expense of shipping such a coffin so far would have been enormous."

"Yes," I replied, "but Joseph was a wealthy man. He could have afforded it."

"I suppose, I suppose. But even if it is true, it must remain conjecture. After two thousand years, buried anywhere in thousands of square kilometers of sand, the body of Jesus will have returned to the earth long ago. We'll simply never know. So, my friend, you now have another mystery to haunt you!"

I remained silent. The strange look that must have been upon my face no doubt was interpreted by the professor as disappointment. But it was nearly uncontrolled thrill. Because I knew the professor was wrong.

He was wrong because I knew the recent history of the Temple of the Four Kings, while he was thinking only in terms of the past. The temple is more correctly called the Temple of Abu Simbel in Nubia, about 1,100 km south of Cairo at the archaeological site of Gebel Adda. In 1960, when Nasser announced the plans for the Aswan High Dam, it was immediately recognized that the resulting floodwaters would drown Gebel Adda forever. A hurried salvage operation was thus started, desperately trying to save what could be saved in the short time left. Among the artifacts recovered were more than 5,000 human skeletons, and several ancient coffins, all of which were hurriedly cataloged and shipped to Cairo. The skeletons have since been extensively studied for bone and dental evolution.

But not the mummies. Considered as just more Egyptian mummies among many already carelessly scattered in a back room of the museum's second-floor gallery room, they had been mostly ignored, as it was standard policy of the museum not to unwrap any mummy unless such a procedure was part of an ongoing scholarly study. But all incoming mummies *were* subjected to a routine X-ray scan, which was then filed. On my arrival at the museum, I had been allowed to go through these archived scan pictures as part of my orientation. One set had briefly caught my attention. The young woman in the records office had been unconcerned, however, about the problem.

"Miss," I recall saying, "this group of pictures is foggy. It almost looks like an underexposure."

"Oh," she replied, glancing quickly at the file in my hands, "it looks more like the plates weren't properly aligned. The X-ray gun is placed on one side of the coffin, you know, and the film on the other, and —"

"Yes, yes, thank you. I am familiar with the technique." I put the file back and forgot it. The explanation seemed perfectly plausible at the time.

But as I sat silent before the professor I *knew* what the real answer was. The film had been aligned properly, all right. But the mummy inside was wrapped in lead! Just as that ancient letter by Joseph of Arimathea had said. The body of Jesus Christ was inside that coffin—I was sure of it! —resting in a dusty storage room not more than 100 m from my office!

How can I convey to you the excitement that charged my mind? I knew I had found the ultimate link between the modern world and the world of a man who had changed the course of history. There *is* no way for you to understand—I was like one overwhelmed by passion. I was to be the first man in twenty centuries to gaze once more upon the features of the Messiah, Jesus of Nazareth.

The arrogance, the blasphemy of that desire, shames me now as I sit here in America recalling that incredible instant of revelation. But at that moment the lure was irresistible. It had to be done in secret, of course—how could I possibly go to the museum authorities as an outsider, an American on a temporary visa, and tell them they had Jesus Christ in their storage room? They would quite properly have had me locked up for observation. No, I had to do it alone.

A week after receiving Jack's letter, under the pretense of working late, by the stroke of midnight I was the sole inhabitant of the second-floor gallery. Armed with a crowbar, a flashlight, and a heavy scissors, I made my way to the storage room. My heart was about to burst, my mind was reeling. I felt it was the greatest moment of my life.

I had looked up the catalog number of the coffin with the foggy X-ray plates and, after about 20 min, I found it. I was in a dark corner of the room, covered with old packing crate materials and a layer of dust a few centimeters thick. It hadn't been touched in decades. I soon had the coffin cleared away and began to pry the lid off with the crowbar. It was sealed solid with the ages, but my wild excitement gave me the strength of ten men. I had the lid off!

Before me lay a mummy, wrapped in the usual bandages, apparently no different from any of the dozens of other mummies scattered about the museum. I picked up the scissors. But then I looked more closely. In the dim light of the dusty corner, I had at first failed to notice that the bandages were charred, almost burned. And they had a dull-gray glint to them, as if they had been sprayed with metallic paint. As I gazed in wonderment at the strange sight, the scissors slipped unnoticed from my hand. In awe, I stood frozen, unable to move. I felt I was in the presence of something that should be left untouched. But then, using the flashlight, I spotted an object lying free in the coffin. As I gazed at what I retrieved, a deep feeling of intrusion swept over me again even more intensely.

At the foot of the ancient sarcophagus, nearly out of sight, I had found something that Joseph had failed to mention in his letter. The Roman procurator of Judea, Pontius Pilate, as a mocking thrust at the chief priests who had condemned Jesus, had had a placard bearing the charge against him nailed to the Cross. Joseph had removed the notice from the Cross, along with Jesus, and placed it in the coffin to bear witness to the identity of the man it accompanied into eternity. Since it is written threefold, in Greek, Latin, and Hebrew, there is little doubt in my mind whose blood it was that still lies splattered across the ancient sign:

JESUS THE NAZARENE
KING OF THE JEWS

I slowly replaced the lid, nearly overcome with emotion. Had I gone too far, pushed scientific curiosity beyond reason into a region where it had no business intruding? I returned to my office, carrying the bloody sign, wrapped in my agonized thoughts.

But still, I couldn't let it be. I recall I stared at the sign for hours, there in the gloomy quiet of my office. I *had* to learn its secrets.

The blood. It is all that is left, the only physical remains of Christ's body that I had—except for the mummy, which I didn't have the courage to free from its wrappings. Once again I wrote to America, this time to ask a friend in the pathology laboratory of the medical school at my university, an expert in forensic medicine, to run a total blood analysis on a fragment of wood I 'happened' to have. Of course, I told him nothing of its origin! I don't know what I expected to learn, but I couldn't help myself. I was so close to understanding the mystery that had haunted me for so long (or so I thought) that I just couldn't leave it alone. And the blood was all I had.

I soon had my friend's report. Halfway through it I had to put it down in shock. It couldn't be! But if it is so, then what could God have intended by allowing His Son to be so cursed? The lab analysis stated the presence of a large number of distinctly malformed lymphocytic cells, absolute, positive diagnosis of lymphosarcoma. Today we'd treat it with amethopterin and 6-mercaptopurine, with the usual result being total remission, possibly lasting as long as 5–7 years. Two thousand years ago, however, it would have been a virtual death sentence, with a survival time of mere weeks.

If Jesus had not died on the Cross, He'd have been dead within 2 months anyway. Or so I thought when I read the analysis. Such a death would have destroyed the perception of Him as the Messiah in the eyes of His followers. So, in that sense, the Crucifixion came just in time. But why the acute leukemia? Why a disease of imperfect man in the body of the Son of God?

The last half of the blood analysis had a second surprise for me, one presenting a riddle that in its own way was even more profound than the first. My friend's delicate chemical tests had also detected the presence of certain toxic blood reaction products—end-stage products produced only by the synthetic drug treatment for leukemia, 6-mercaptopurine, as if He had been on chemotherapy and was beginning to suffer a relapse just before His execution.

I sat stunned, numb with disbelief. It was all so incredible. I pulled open my bottom desk drawer, the one I always kept locked since hiding the sign away in it. As I held the old wood in my hands, I began to doubt. Was the sign really old, or was it all just a fantastic hoax? Was Professor el-Zam merely making me the butt of an elaborate, cunning, terrible prank? I knew then what I had to do, what I must do, if I was to know another moment free from confusion.

I had to open the mummy!

The very next night found me again in the storage room. I was now almost in a fever pitch of excitement and had the coffin lid off in just a minute or two. I attacked the oddly metallic, burnt bandages with my scissors. But what!—the ancient cloth fell apart at the thrust of the blades. What revealed itself to me was so astounding I dropped the scissors and staggered backwards. When I had begun to cut, the cloth had been of the form of a heavily wrapped figure. But as it separated under the force of the blades it fell in on itself, as if it contained nothing but space. And indeed there were no bones, nothing to mark the resting place of a man. But there *was* something there. A congealed, roughly spherical ball of lead!

It had to be the lead sheets, the soft lead foil mentioned in Joseph's letter. I stood paralyzed by both wonderment and surprise. There was no doubt now, this *was* the coffin of Jesus. But—where was He?

Slowly, the connection between the charred bandages and the melted lead sheets became clear in my bewildered mind. Something had caused the temperature of the covered body to reach at least the temperature of molten lead—328 °C. If the heating had occurred in just the right way, the surface tension of the liquid metal would have pulled it into a sphere. And lead mist would have impregnated the charred bandages, causing the fogging on the x-ray plates that I saw. I thought to myself: *It would have taken energy, wouldn't it—a lot of energy—for Him to return to His Father!*

The mystery that had bedeviled me for so long was finally resolved. The bloody sign, with its incredible tale, and the ancient coffin, had shown me the way. When dying on the Cross, Jesus had cried out in agony, "My God, my God, why hast thou forsaken me?" If only He could have known then, in His moment of extreme anguish, that His Father had *not* abandoned Him, but had given Him the gift of life beyond His natural time on Earth.

As my thoughts dwell on my discoveries, and as I think of the compassion of the Lord God Almighty for His Son, I feel comfort and warmth. I feel at real peace with myself for the first time in my life. I didn't do wrong in pursuing the riddle.

I *have* found Christ.

****

If you find "A Father's Gift" to be borderline outrageous (which was, in fact, my goal when I wrote it—as well as projecting, just a bit, the excitement of a 'scholarly hunt'), well then, your brain will positively overheat and perhaps even *melt* if you read a book written decades later by authors who claim they have found the Tomb of the entire Jesus clan: *The Jesus Family Tomb: the discovery, the investigation, and the evidence that could change history* by Simcha Jacobovici and Charles Pellegrino, HarperCollins 2007. The well-known film producer and director James Cameron, of *Titanic* fame, wrote a quite literate Forward to the book and, since he once lived in the same small Southern California town that I grew-up in (Brea), I was initially inclined to give the book a good read. Yes, I know, not a lot of correlation there, and so you should read it and make-up your own mind about its merits.

Appendix 4: "Applied Mathematical Theology"

Published in *NATURE*, March 2, 2006. Copyright 2006 by Gregory Benford (reprinted by permission of the author)

The discovery that the Cosmic Microwave Background has a pattern buried in it unsettled the entire world.

The temperature of this 2.7 K. emission left over from the Big Bang, varies across the sky. Temperature ripples can be broken into angular- coordinate Fourier components, and this is where radio astronomers found a curious pattern—a message, or at least, a pattern. Spread across the microwave sky there was room in the detectable fluctuations for about 100,000 bits—roughly 10,000 words.

Although different technical civilizations in our Universe would see different temperature fluctuations, they could agree on the Fourier coefficients. This independence of place, and the role of the cosmic background as cosmic neon sign for anyone with a microwave receiver, meant that any intelligence in the Universe could see this pattern.

But what did it mean? Certainly it would not be in English or any other human language. The only candidate tongue was mathematics.

Writing them as binary numbers, astronomers tried to fit mathematical sequences, such as the prime numbers, in any base. This and other mathematical favorites—pi, e, the golden ratio, the Riemann zeta function—proved futile. More obscure numbers and patterns, from set theory and the like, also shed no light.

In despair, some thought the pattern might be random. But the Shannon entropy test showed clear non-random elements, and this nihilist idea faded away. One insight from Benford's Law, which states that the logarithms of artificial numbers are uniformly distributed, did apply to the tiny fluctuations.

P.J. Nahin, *Holy Sci-Fi!*, Science and Fiction,
DOI 10.1007/978-1-4939-0618-5, © Springer Science+Business Media New York 2014

This proved that the primordial microwaves were not random, and so had been artificially encoded, perhaps by some even earlier process. So there was a message, of sorts.

Cosmologists eagerly searched for clues and hit a dead end. The sequence was found to fit no model. This suggested immediately to even nonreligious astronomers that the pattern may have been put there by a being who made our universe: God, in short.

What would such a mathematical message mean, anyway? Only that some rational, counting designer had made our Universe. Beyond that, nothing would be revealed about the being's nature; though of course it would prove the old claim, that God was a mathematician.

Rankled, the physicists quickly compared the observed sequence with the fine structure constant, one of their favorites. The sequence did not fit.

This sent everyone back to fundamentals. Current theory says that tiny temperature fluctuations in the microwaves came from little bumps in the potential function that governed the inflation of the very early universe. Tinkering with those quantum fluctuations, a being could write something simple but profound: God as a quantum mechanic. If, for example, the designer could encode little squiggles on the potential, then the fine-tuned primordial density fluctuations would not be exactly scale-free, and that's where the sky-wide microwave patterns came from.

So of course the physicists followed their current fashion. When comparison with other favorite numbers—the dimensionless ratios of masses and energies and the like—all failed, they tried more advanced theories. They tried prescriptions for various symmetry groups that came from the Lie algebras, as three of the four fundamental interactions we know reflect such gauge theories. No help.

The physicists, who had long been the mandarins of science, then supposed that clues to the correct string theory, a menu currently offering about 10^{100} choices, would be the most profound of messages. After all, wouldn't God want to make life easier for physicists? Because, obviously, God was one, too.

Sadly, no. Nothing seemed to work.

Perhaps the very idea underpinning science—that humans could understand the Universe—had hit a wall. This helped both science and religion.

Excitement increased. If the being was not saying something obvious, then maybe humans had not understood the Universe enough to make out the message. Governments poured money into mathematics and physics. The astronomers protested. If the night sky was a tale told by God, they could read it. The cosmic neutrino and gravity wave backgrounds had not yet been detected, but they could also carry the Word. So it came to be that the cosmologists, too, received the blessing of a large research bounty.

These huge increases in funding drove a renaissance of modern science. Data processers, statistical theorists, observers of obscure spectra—all received a shared. Vast telescopes tuned to the vibrations and emissions of the Universe glided in high orbits, their ears cupped to the distant and primordial.

This largess produced an economic boon, too, as many spinoff technologies benefited commerce. Religious fervor damped, as each faith felt humbled by this proof that the Universe had meaning, yet mankind was not yet advanced enough to fathom it.

At the same time, attention focused on the injunction to mankind in the Old Testament—echoed in other religious founding texts—charging humanity with being the stewards of Earth. The environmental movement merged with the great religions.

Within a century, active adjustment of Earth's reflected sunlight, and capturing of carbon in the oceans and lands, had averted the greenhouse disaster. Church attendance was enormous. Efforts to enhance our knowledge and skills had averted many gathering social conflicts.

Work on the Message continues in the new university departments of Applied Mathematical Theology. Yet to this day, the Message remains untranslated. Perhaps that is just as well.

Appendix 5: "Gravity's Whispers"

"The best is the enemy of the good," Sam said over my shoulder.

I whirled around, knowing the voice, smiling. "What —?"

He sauntered in, grinning in his lopsided way. "At 11 p.m. you're still working. Know your limits. The data can't get better when you're tired, y'know."

I threw down my pencil. "Right. Pursue the good. Let's get a beer."

At the Very Large Array, this meant a long drive back to Socorro. Our offices were there, but I liked spending time out among the big radio dishes, too. On the way back I rolled down the window to smell the tangy spring sagebrush and wondered whether Sam the Slow had finally decided to make a date with me, in his odd way. I'd been waiting half a year.

Then he said: "I was just passing by, thought I'd follow up on that puzzle I sent last week."

He had sent through a noise-dominated file. I had run one of my custom programs, gotten interested, and wasted a day pulling out a pattern. "You know me too well. I cracked it, yeah." I gave him a smile he didn't notice. "Not a very interesting solution."

"You'd be surprised," Sam said, watching the desert slide by.

"It's you guys who surprised the world—the first gravity waves, wow."

"Yeah, decades of work on LIGO paid off."

Sam was also modest, a trait that gave him gal problems in the fanatic tech crowd more than once. Getting a gravitational wave to tweak a cavity, and detect that with interfering waves, had burned 20 years of his life. He

P.J. Nahin, *Holy Sci-Fi!*, Science and Fiction,
DOI 10.1007/978-1-4939-0618-5, © Springer Science+Business Media New York 2014

shrugged. "We thought it was a signal from a rotating neutron star with a deformed crust. Say, you have that solution handy?"

I flipped open my laptop. "It's a string of numbers, turns out to be the zeroes of the Riemann zeta function."

"Uh huh. Which is —?"

"A famous function of complex argument. It analytically continues the sum of an infinite series."

"Sounds boring."

"Not so." At least he was looking at me now. "It's a big deal in analytic number theory, plenty of applications in physics, probability theory, Bose–Einstein condensates, spin waves —"

"Useful, good." Sam was usually sharp, focused, but now he gazed pensively at the stars.

"So how'd you get the detection?" It would help if I got him started about his work —that is, his life. "You guys got rid of the noise from that road traffic and logging at the Louisiana site?"

"Yeah, took years. The signal we finally got had plenty of chirps and bursts in it, a bitch to clean up."

I grinned. Sam had worked decades on LIGO, and now the milestone was here. "Now that you've got LIGO sensitive enough, there'll be plenty of signals. Supernovas in other galaxies, maybe rattling cosmic strings —"

"I want to understand this one. It's not a neutron star crust vibration, I think."

"Huh?" I was already tasting the beer in my mind.

"That decoding you did? That was our signal."

I blinked. "Can't be. No natural system—"

"Exactly." Sam hooked an eyebrow at me.

"What? A tunable gravitational wave with a signal? That's im—"

"—possible, I know. Unless you can sling around neutron stars and make them sing in code."

Maybe, just maybe, this could be more important than at last getting Sam to date me. Maybe. "Then... you should know that it's not just a list of numbers. After 20 of the Riemann zeros, there's something like a proof of the Riemann hypothesis."

He frowned. "Uh, so?"

"It's one of the greatest unsolved problems in mathematics. It says that any non-trivial zero has its real part exactly equal to 1/2."

He shook his head. "And that's the attention-catching opener to a SETI signal?"

"So you see, it can't be. Opening up with pi, or e, prime numbers, the fine structure constant—that makes sense."

"Sense to the likes of us."

"So I must've made some mistake."

"No you didn't." Sam looked at me with a warm smile. "You're the only one I could run to with this analysis—the rest of 'em would laugh. You're good, really good."

I leaned over and kissed him. "Congratulations on the Nobel."

He kissed back, his eyes flickered, he grinned—but he didn't look happy. He grasped the steering wheel and peered ahead into the starlit darkness. In the high desert you can see stars above the headlights. I knew him enough to see that he was thinking about something that could whisper across the galaxies with gravitation, not using obvious means like radio or lasers. "Any mind that thinks the Riemann numbers are a calling card—and can throw around stars. . ."

I got it. "Yeah. Know your limits. Maybe it's good, really good, that we can't possibly answer them."

****

Two elaborative comments on references in "Gravity's Whispers": (a) LIGO is a real scientific effort (with substantial funding, at the level of several hundreds of millions of dollars, from the National Science Foundation,), designed to detect gravitational waves. LIGO stands for the *Laser Interferometer Gravitational-Wave Observatory:* it is a joint project staffed with scientists from MIT, Caltech, and other colleges and universities. (b) You can find more about the Riemann hypothesis in my book *An Imaginary Tale: the story of* $\sqrt{-1}$, Princeton University Press 2010, pp. 150–155.

Bibliography of Short Stories Cited[7]

Anderson, Poul, "The Problem of Pain." *Chronicles of a Comer: and other religious science fiction stories* (Roger Elwood, editor, John Knox Press 1974.

—, "The Word to Space," *Other Worlds, Other Gods* (Mayo Mohs, editor), Avon 1971 (published under the pen-name of Winston P. Sanders).

—, "A Chapter of Revelation," *The Day the Sun Stood Still* (Lester del Rey, editor), Thomas Nelson 1972.

Asimov, Isaac, "The Bicentennial Man" and "Evidence," *Machines That Think* (Isaac Asimov, *et al.*, editors), Holt, Rinehart and Winston 1983.

—, "The Last Question" and "The Last Answer," *Robot Dreams*, Ace 1986.

—, "Reason," *Science Fiction: a historical anthology* (Eric S. Rabkin, editor), Oxford University Press 1983.

—, "The Tercentenary Incident," *The Bicentennial Man and Other Stories*, Doubleday 1976.

—, "That Thou Art Mindful of Him," *Souls in Metal* (Mike Ashley, editor), St. Martin's Press 1977.

—, "The Ugly Little Boy," *Creations: the quest for origins in story and science*, (Isaac Asimov, *et al.*, editors), Crown 1983.

---

[7] In this section I've included only short stories (and not book-length works) as they are by far the more difficult to locate. To find a copy of Dante's *Inferno*, or of Arthur C. Clarke's *2001: A Space Odyssey*, or of Carl Sagan's *Contact*, or of H. G. Wells' *War of the Worlds*, shouldn't really be very difficult. The stories listed are most easily found today reprinted in anthologies, and the really good stories are in multiple anthologies. The specific anthology I've given here for each story simply happens to be the one I first used. In the case of short stories that I couldn't find in an anthology, I've immediately cited in the text their original (and only, I suspect) magazine appearance. An older but still quite useful book with an extensive bibliography containing numerous theological SF stories that I have not discussed in this book is *The Transcendent Adventure: studies in religion in science fiction/fantasy* (Robert Reilly, editor), Greenwood Press 1985. I certainly make no claim to have included every last SF story with a religious theme. Many readers will no doubt have a story in mind about which he or she will ask, 'Why isn't [insert your favorite title] here?' The easy answer is, 'I didn't read it' or, perhaps, I *did* read it and decided its message had already been discussed in some other tale. The historian's inescapable fate, I fear, is to be second-guessed by well-read readers! You can write to me at paul.nahin@unh.edu if you want to tell me what you think should be included in a later edition.

P.J. Nahin, *Holy Sci-Fi!*, Science and Fiction,
DOI 10.1007/978-1-4939-0618-5, © Springer Science+Business Media New York 2014

Asimov, Isaac, "Trends," *The Early Asimov, or Eleven Years of Trying*, Doubleday 1972.

Balchin, Nigel, "God and the Machine," *Fantasia Mathematica* (Clifton Fadiman, editor), Simon and Schuster 1958.

Beaumont, Charles, "Last Rites," *The Magic Man and Other Science-Fantasy Stories*, Fawcett 1965.

Benford, Gregory, "Anomalies," *Redshift: extreme visions of speculative fiction* (Al Sarrantonio, editor), ROC 2001.

Bierce, Ambrose, "Moxon's Monster," *Machines That Think* (Isaac Asimov, *et al.*, editors), Holt, Rinehart and Winston 1983.

Binder, Eando, "I, Robot," *The Coming of the Robots* (Sam Moskowitz, editor), Collier Books 1963.

Bond, Nelson, "The Cunning of the Beast," *Other Worlds, Other Gods* (Mayo Mohs, editor), Avon 1971.

—, "Uncommon Castaway," *No Time Like the Future*, Avon 1954.

Boucher, Anthony, "The Quest for Saint Aquin," *Sacred Visions* (Andrew M. Greeley and Michael Cassutt, editors), TOR 1991.

Bova, Ben, "Inspiration," *Holt Anthology of Science Fiction*, Holt, Rinehart and Winston, 2000.

Bradbury, Ray, "The Man" and "The Fire Balloons," *The Illustrated Man*, Doubleday 1951.

Brunner, John, "The Vitanuls," *Other Worlds, Other Gods* (Mayo Mohs, editor), Avon 1971.

—, "Judas," *Machines That Think* (Isaac Asimov, *et al.*, editors), Holt, Rinehart and Winston 1983.

—, "The Windows of Heaven," *Yet More Penguin Science Fiction* (Brian Aldiss, editor), Penguin 1964.

Campbell, Jr., John W., "Who Goes There?" and "The Last Evolution," *The Best of John W. Campbell* (Lester del Rey, editor), Nelson Doubleday 1967.

Chiang, Ted, "Hell is the Absence of God," *The Locus Awards* (Charles N. Brown and Jonathan Strahan, editors), HarperCollins 2004.

Clarke, Arthur C., "The Nine Billion Names of God," *Other Worlds, Other Gods* (Mayo Mohs, editor), Avon 1971.

—, "The Star," *Science Fiction: a historical anthology* (Eric S. Rabkin, editor), Oxford University Press 1983.

—, "Dial F for Frankenstein" and "The Sentinel," *The Collected Stories of Arthur C. Clarke*, TOR 2000.

Davidson, Avram, "Or the Grasses Grow," *Science Fiction Showcase* (Mary Kornbluth, editor), Mayflower 1968.

Del Rey, Lester, "For I Am a Jealous People," *The Best of Lester del Rey*, Ballantine 1978.

—, "Evensong," *Other Worlds, Other Gods* (Mayo Mohs, editor), Avon 1971.

—, "Helen O'Loy," *Machines That Think* (Isaac Asimov, *et al.*, editors), Holt, Rinehart and Winston 1983.

Dick, Philip K., "Jon's World," *Time to Come* (August Derleth, editor), Farrar, Straus and Young 1954.

—, "The Skull" and "The Great C," *Beyond Lies the Wub*, Gollancz 1988.

Dickson, Gordon R., "Things Which Are Caesar's," *The Day the Sun Stood Still* (Lester del Rey, editor), Thomas Nelson 1972.

Ellison, Harlan, "I Have No Mouth & I Must Scream," *I Have No Mouth & I Must Scream* Edgeworks Abbey 2009.

Farmer, Philip José, "Riverworld," *Down in the Black Gang*, Nelson Doubleday 1971.

Fitzgerald, F. Scott, "The Curious Case of Benjamin Button," *Pause to Wonder*, J. Messner 1944.

Gallun, Raymond Z., "Derelict," *The Coming of the Robots* (Sam Moskowitz, editor), Collier Books 1963.

Gernsback, Hugo, "Ralph 124C 41+," *Science Fiction: a historical anthology* (Eric S. Rabkin, editor), Oxford University Press 1983.

Godwin, Tom, "The Cold Equations," *The Cold Equations and Other Stories* (Eric Flint, editor) Baen 2004.

Gunn, James E., "Kindergarten," *Creations: the quest for origins in story and science*, (Isaac Asimov, *et al.*, editors), Crown 1983.

Harness, Charles, "Child by Chronos," *The Best from Fantasy and Science Fiction* (volume 3), Doubleday 1954.

Heinlein, Robert, "All You Zombies —," *The Mirror of Infinity* (Robert Silverberg, editor), Canfield 1970.

—, "By His Bootstraps," *The Arbor House Treasury of Great Science Fiction Short Novels*, Arbor House 1980.

—, "Universe" and "Common Sense," *Orphans of the Sky*, Baen 2001.

Keizer, Gregg, "Angel of the Sixth Circle," *Perpetual Light* (Alan Ryan, editor), Warner 1982.

Kilworth, Garry, "Let's Go to Golgotha!," *The Songbirds of Pain*, Victor Gollancz 1984.

Knight, Damon, "Shall the Dust Praise Thee?" *Dangerous Visions* (Harlan Ellison, editor), Simon and Schuster 1967.

Leinster, Murray, "First Contact," *Aliens* (Ben Bova, editor), St. Martin's Press 1977.

—, "Sidewise in Time," *Before the Golden Age: a science fiction anthology of the 1930s*, (Isaac Asimov, editor), Doubleday 1974.

Lem, Stanislaw, "Project Genesis," *Creations: the quest for origins in story and science*, (Isaac Asimov, *et al.*, editors), Crown 1983.

—, "Non Serviam," *A Perfect Vacuum*, Northwestern University Press 1999.

Lewis, Anthony R., "The Turing Test," *Deals with the Devil* (Mike Resnicl *et al.*, editors) Daw 1994.

Martin, George R. R., "The Way of Cross and Dragon," *The Locus Awards* (Charles N. Brown and Jonathan Strahan, editors), HarperCollins 2004.

Matheson, Richard, "The Traveller," *Third from the Sun*, Bantam 1954.

McDevitt, Jack, "Friends in High Places," *A Cross of Centuries* (Michael Bishop, editor), Thunder's Mountain Press 2007.

McDevitt, Jack, "Time's Arrow," *The Fantastic Civil War* (F. McSherry, Jr., editor), Baen 1991.

Merritt, Abraham, "The Last Poet and the Robots," *Science Fiction: a historical anthology* (Eric S. Rabkin, editor), Oxford University Press 1983.

Moorcock, Michael, *Behold the Man*, *A Cross of Centuries* (Michael Bishop, editor), Thunder's Mountain Press 2007.

Payes, Rachel Cosgrove, "In His Own Image," *Strange Gods* (Roger Elwood, editor), Pocket 1974.

Porges, Arthur, "The Devil and Simon Flagg," *Fantasia Mathematica* (Clifton Fadiman, editor), Simon and Schuster 1958.

—, "The Rescuer," *Yet More Penguin Science Fiction* (Brian Aldiss, editor), Penguin 1964.

Russell, Erik Frank, "Second Genesis," *Deep Space*, Fantasy Press 1954.

Schenck, Hilbert, "The Theology of Water," *Perpetual Light* (Alan Ryan, editor), Warner 1982.

Sheckley, Robert, "Can You Feel Anything When I Do This?" *A Science Fiction Argosy* (Damon Knight, editor), Simon and Schuster 1972.

Silverberg, Robert, "Good News from the Vatican," *Beyond the Safe Zone*, Donald I. Fine, Inc., 1986.

—, "The Assassin," *101 Science Fiction Stories* (Martin Greenberg and Charles Waugh, editors), Avenel 1986.

—, "Thomas the Proclaimer," *The Day the Sun Stood Still* (Lester del Rey, editor), Thomas Nelson 1972.

Spinrad, Norman, "The Weed of Time," *Alchemy & Academe* (Anne McCaffrey, editor), Ballantine 1980.

Sturgeon, Theodore, *"Microcosmic God,"* *Microcosmic God: the complete stories of Theodore Sturgeon* (volume 2), North Atlantic Books 2010.

Tenn, William, "On Venus, Have We Got a Rabbi," *Wandering Stars: an anthology of Jewish Fantasy & Science Fiction* (Jack Dann, editor), Jewish Lights Publishing 1998.

Thomas, John B., "Return to a Hostile Planet," *Strange Gods* (Roger Elwood, editor), Pocket 1974.

Van Vogt, A. E., "The Seesaw," *Creations: the quest for origins in story and science*, (Isaac Asimov, *et al.*, editors), Crown 1983.

Vincent, Harl, "Rex," *The Coming of the Robot* (Sam Moskowitz, editor), Collier Books 1963.

Wells, H. G., "Davidson's Eyes" and "The Plattner Story," *Best Science Fiction of H. G. Wells*, Dover 1966.

—, "The Star," *Science Fiction: a historical anthology* (Eric S. Rabkin, editor), Oxford University Press 1983.

Wilcox, Don, "The Voyage that Lasted 600 Years," *Isaac Asimov Presents the Best Science Fiction Firsts*, Barnes and Noble Books 1996.

Wyndham, John, "The Lost Machine," *Machines That Think* (Isaac Asimov, *et al.*, editors), Holt, Rinehart and Winston 1983.

Index

P.J. Nahin, *Holy Sci-Fi!*, Science and Fiction,
DOI 10.1007/978-1-4939-0618-5, © Springer Science+Business Media New York 2014

Printed by Printforce, the Netherlands